INTERNATIONAL AND METRIC
UNITS OF MEASUREMENT

by

MARVIN H. GREEN

*Supervisor of Research
and Development
New Jersey State Bureau
of Air Polution Control*

CHEMICAL PUBLISHING CO., INC.

1973

International and Metric
Units of Measurement

ISBN: 978-0-8206-0150-2

Chemical Publishing Company:
www.chemical-publishing.com
www.chemicalpublishing.net

First Edition: Chemical Publishing New York 1973

Printed in the United States of America

TABLE OF CONTENTS

FOREWORD

In the *National Bureau of Standards Technical News Bulletin,* **43,** No. 1 (January 1959), it was announced that the international yard and pound would be adopted, effective July 1, 1959, as agreed upon by the directors of the following standards laboratories:

Applied Physics Division, National Research Council, Ottawa (Canada)

Dominion Physical Laboratory, Lower Hutt (New Zealand)

National Bureau of Standards, Washington (United States of America)

National Physical Laboratory, Teddington (United Kingdom)

National Physical Research Laboratory, Pretoria (South Africa)

National Standards Laboratory, Sydney (Australia)

The international units have the following definitions:

the international yard = 0.9144 meter
the international pound = 0.45359237 kilogram

The international inch derived from the international yard is exactly equal to 25.4 millimeters. This value for the inch had been adopted by Canada, the American Standards Association in 1933, the National Advisory Committee for Aeronautics in 1952, and many standardizing organizations in other countries.

Thus after many years of consideration, the units based upon the Mendenhall Order, which served as the basis for United States customary units, have been replaced by the international units. In the Mendenhall Order, adopted in 1893, the yard is defined as exactly 3600/3937 meter. From this definition, the meter equals 39.37 inches and the inch then approximates 25.4000508 millimeters. The inch used by the National Physical Laboratory of the United Kingdom equals 25.399956 millimeters.

The pound units replaced by the international pound are as follows:

1 United States pound = 0.4535924277 kilogram
1 British pound = 0.453592338 kilogram
1 Canadian pound = 0.45359243 kilogram

The international pound of 0.45359237 kilogram was selected so as to be exactly divisible by 7 to give the following exact value for the grain:

$$1 \text{ international grain} = 0.06479891 \text{ gram}$$

The international inch is approximately 2 parts per million shorter than the United States customary inch. The value for the meter has not been altered.

In the *Technical News Bulletin,* **44,** No. 12 (December 1960), it was announced that the world had adopted a new international standard of length on October 14, 1960. The action by the Eleventh General Conference on Weights and Measures replaces the meter bar which had served as the standard for length since 1889 under the Treaty of the Meter. Originally the meter was defined as a unit of length equal to the distance between the defining lines on the International Prototype Meter when this standard is the temperature of melting ice (0° C). The new definition of the meter is as follows:

> One meter is equal to 1,650,763.73 wave lengths of the orange-red line of krypton 86.

The new definition of the meter will not materially change the measurement of length nor in any way alter the relationship between the English (or international) and metric units. The inch is equal to 41,929.399 wave lengths of the orange-red line of krypton 86, based on one inch being equal to 2.54 centimeters.

The Conference also confirmed the action of the International Committee on Weights and Measures in defining the second of time as 1/31,556,925.9747 of the tropical year 1900. In addition, the Conference adopted the symbols SI (Système Internationale) to designate the system of units involving the meter, kilogram, second, and related units and named the unit of magnetic flux density in this system the "tesla".

The international pound is approximately 1-1/2 parts in ten million smaller than the United States customary pound, and other mass units bear the same relationship to their previously used counterparts. The definition of the kilogram remains as defined in *National Bureau of Standards Miscellaneous Publication 233:*

A kilogram is a unit of mass equal to the mass of the International Prototype Kilogram.

It is not likely that the definition of the kilogram will be altered in the near future.

Appendix B lists the comparisons of the international and United States customary units for many of the sections in this book. The international units of volume are approximately 6 parts per million smaller than the United States customary units, and a density measurement based on international units of mass and volume is almost 6 parts per million larger than one based on United States customary units.

Throughout the book, international units have replaced customary units. For example, the horsepower is equal to 550 international foot pounds per second, and is therefore approximately 4 parts per million smaller than the horsepower based on the customary pound and foot.

Calculations are based wherever possible upon exact values rather than on possibly inexact reciprocals. Results of calculations are usually carried out to nine or ten figures and then rounded off to seven significant figures for listing in the.tables. Throughout the book units are listed in order of increasing magnitude in the various tables, including those listing additional units in the measurement categories. All calculations were thoroughly checked and cross-checked to assure accuracy.

INTRODUCTION TO THE SECOND EDITION

One of the major characteristics of *"International and Metric Units of Measurement"* is extreme accuracy. In order to retain this accuracy it was necessary to up-date many sections of the book because of changes that have been made in recent years in the field of measurement.

One important change is a new definition of the liter. It is now equal to 1000 cubic centimeters and is no longer related to a volume of water as under the previous definition. Although the new liter is only 28 parts in a million smaller than the old unit, this small change necessitated revisions in three sections of the book: volume, flow, and density and concentration. International Steam Table units, the gram calorie and the related British Thermal Unit, have new values, and in this case it was necessary to revise the sections on energy and power units. The section on pressure units has also been up-dated with the atmosphere taken as equal to 1,013,250 dynes per square centimeter, a very small change from the previous definition of the atmosphere as equal to 760 millimeters of mercury at $0°$ C. A new value for the speed of light in a vacuum, universally known as "c", made it necessary to revise three more sections: atomic energy, electrical, and magnetic units. A more recent value of the Avogadro constant "N" produced additional changes in the tables of atomic energy units.

Nine of the sixteen sections of the book have been revised. In three more sections the text was altered to include the most recent definitions of fundamental units of measurement. The three fundamental units of measurement are the meter in the section on length, the kilogram in the section on mass, and the second in the section on time. Of the three only the second has been completely re-defined on the basis of the vibrations of the cesium-133 atom.

It is intended that the second edition serve as a more up-to-date compilation of units of measurement and their conversion factors.

1. ANGULAR MEASURE

Units of angular measure listed in the table are:

1 second
1 minute = 60 seconds
1 grade = 54 minutes
1 degree = 60 minutes
1 radian = $1/2\pi$ circle
1 quadrant = 100 grades
1 circle = 360 degrees

The relationship between the radian and the circle involves the factor π, which is equal to 3.14159265. By dividing 360 by 2π, the radian is equal to 57.29578 degrees.

The circle, also referred to as the circumference or revolution, is equal to 12 signs, one sign being equal to 30 degrees. The grade is equal to 100 centesimal minutes.

TABLE 1-1

UNITS OF ANGULAR MEASURE	Second	Minute	Grade	Degree	Radian	Quadrant	Circle
Second	1	0.01666667	3.086420×10^{-4}	2.777778×10^{-4}	4.848137×10^{-6}	3.086420×10^{-6}	7.716049×10^{-7}
Minute	60	1	0.01851852	0.01666667	2.908882×10^{-4}	1.851852×10^{-4}	4.629630×10^{-5}
Grade	3240	54	1	0.9	0.01570796	0.01	0.0025
Degree	3600	60	1.111111	1	0.01745329	0.01111111	2.777778×10^{-3}
Radian	206,264.8	3437.747	63.66198	57.29578	1	0.6366198	0.1591549
Quadrant	324,000	5400	100	90	1.570796	1	0.25
Circle	1,296,000	21,600	400	360	6.283185	4	1

2. AREA

Units of area listed in the tables are:

INTERNATIONAL

1 square inch
1 square link = 62.7264 square inches
1 square foot = 144 square inches
1 square yard = 9 square feet
1 square rod = 625 square links
1 square chain = 10,000 square links
1 acre = 10 square chains
1 square mile = 640 acres

METRIC

1 square millimeter = 0.01 square centimeter
1 square centimeter = 0.0001 square meter
1 square meter
1 are = 100 square meters
1 hectare = 100 ares

The two systems of units are related by the following:

1 international square inch = 6.4516 square centimeters

The international units of area are based on the international units of length adopted by the National Bureau of Standards, effective July 1, 1959. The international and old United States customary units of area are related as follows:

1 international unit = 0.999996 U.S. customary unit
1 U.S. customary unit = 1.000004 international units

ADDITIONAL UNITS OF AREA

INTERNATIONAL

circular mil = area of a circle of 1 mil diameter
= 0.7853982 square mil
square mil = 0.000001 square inch

circular inch = area of a circle of 1 inch diameter
= 0.7853982 square inch
Gunter's (surveyor's) square link = 1 square link (in tables)
= 62.7264 square inches
Ramden's (engineer's) square link = 1 square foot
square perch = 1 square rod (in tables) = 272.25 square feet
Gunter's (surveyor's) square chain = 1 square chain (in tables)
= 4356 square feet
Ramden's (engineer's) square chain = 10,000 square feet
section = 1 square mile = 27,878,400 square feet
township = 36 square miles

One acre is equal to a square 69.57011 yards on a side, or 208.7103 feet on a side.

METRIC

barn = 1×10^{-24} square centimeter
circular millimeter = 0.7853982 square millimeter
square decimeter = 0.01 square meter
centare = 0.01 are = 1 square meter
square dekameter = 1 are
square hectometer = 1 hectare
square kilometer = 100 hectares = 1,000,000 square meters

TABLE 2-1

UNITS OF AREA	Sq. Millimeter	Sq. Centimeter	Sq. Inch	Sq. Link	Sq. Foot	Sq. Yard	Sq. Meter
Sq. Millimeter	1	0.01	1.550003×10^{-3}	2.471054×10^{-5}	1.076391×10^{-5}	1.195990×10^{-6}	0.000001
Sq. Centimeter	100	1	0.1550003	2.471054×10^{-3}	1.076391×10^{-3}	1.195990×10^{-4}	0.0001
Sq. Inch	645.16	6.4516	1	0.01594225	6.944444×10^{-3}	7.716049×10^{-4}	6.4516×10^{-4}
Sq. Link	40,468.56	404.6856	62.7264	1	0.4356	0.0484	0.04046856
Sq. Foot	92,903.04	929.0304	144	2.295684	1	0.1111111	0.09290304
Sq. Yard	836,127.36	8361.2736	1296	20.66116	9	1	0.83612736
Sq. Meter	1,000,000	10,000	1550.003	24.71054	10.76391	1.195990	1
Sq. Rod	25,292,852.64	252,928.5264	39,204	625	272.25	30.25	25.29285264
Are	1×10^{8}	1,000,000	155,000.3	2471.054	1076.391	119.5990	100
Sq. Chain	4.046856×10^{8}	4.046856×10^{6}	627,264	10,000	4356	484	404.6856
Acre	4.046856×10^{9}	4.046856×10^{7}	6,272,640	100,000	43,560	4840	4046.856
Hectare	1×10^{10}	1×10^{8}	1.550003×10^{7}	247,105.4	107,639.1	11,959.90	10,000
Sq. Mile	2.589988×10^{12}	2.589988×10^{10}	4.0144896×10^{9}	64,000,000	27,878,400	3,097,600	2.589988×10^{6}

TABLE 2-2

UNITS OF AREA	Sq. Rod	Are	Sq. Chain	Acre	Hectare	Sq. Mile
Sq. Millimeter	3.953686×10^{-8}	1×10^{-8}	2.471054×10^{-9}	2.471054×10^{-10}	1×10^{-10}	3.861022×10^{-13}
Sq. Centimeter	3.953686×10^{-6}	0.000001	2.471054×10^{-7}	2.471054×10^{-8}	1×10^{-8}	3.861022×10^{-11}
Sq. Inch	2.550760×10^{-5}	6.4516×10^{-6}	1.594225×10^{-6}	1.594225×10^{-7}	6.4516×10^{-8}	2.490977×10^{-10}
Sq. Link	0.0016	4.046856×10^{-4}	0.0001	0.00001	4.046856×10^{-6}	1.5625×10^{-8}
Sq. Foot	3.673095×10^{-3}	9.290304×10^{-4}	2.295684×10^{-4}	2.295684×10^{-5}	9.290304×10^{-6}	3.587006×10^{-8}
Sq. Yard	0.03305785	8.3612736×10^{-3}	2.066116×10^{-3}	2.066116×10^{-4}	8.3612736×10^{-5}	3.228306×10^{-7}

(CONTINUED)

TABLE 2-2 (CONTINUED)

UNITS OF AREA	*Sq. Rod*	*Are*	*Sq. Chain*	*Acre*	*Hectare*	*Sq. Mile*
Sq. Meter	0.03953686	0.01	2.471054×10^{-3}	2.471054×10^{-4}	0.0001	3.861022×10^{-7}
Sq. Rod	1	0.2529285264	0.0625	0.00625	$2.529285264 \times 10^{-3}$	9.765625×10^{-6}
Are	3.953686	1	0.2471054	0.02471054	0.01	3.861022×10^{-5}
Sq. Chain	16	4.046856	1	0.1	0.04046856	1.5625×10^{-4}
Acre	160	40.46856	10	1	0.4046856	1.5625×10^{-3}
Hectare	395.3686	100	24.71054	2.471054	1	3.861022×10^{-3}
Sq. Mile	102,400	25,899.88	6400	640	258.9988	1

3. ATOMIC ENERGY UNITS

Units of atomic energy listed in the tables are:

TABLE 1

1 electron volt
1 kiloelectron volt = 1000 electron volts
1 million electron volt = 1,000,000 electron volts
1 atomic mass unit = 931.479 million electron volts
1 billion electron volt = 1000 million electron volts

TABLE 2

1 gram mass energy equivalent
1 pound mass energy equivalent = 453.59237 grams mass energy
equivalent

For comparison purposes the following units of energy from the energy tables are included in the tables of atomic energy units:

TABLE 1

erg
foot pound
gram calorie

TABLE 2

erg
foot poundal
foot pound
kilogram calorie
horsepower hour
absolute kilowatt hour

The foot pound, foot poundal and horsepower hour are defined in terms of the international units adopted by the National Bureau of Standards on July 1, 1959. The pound is the international avoirdupois pound.

Reference 1 is the source of the following physical constants used to develop relationships among the units in this section:

electron charge, e $= 4.80298 \times 10^{-10}$ electrostatic unit of charge

Avogadro constant, N $= 6.02252 \times 10^{23}$ atoms per gram-atom

speed of light in vacuum, c $= 2.997925 \times 10^{10}$ centimeters per second

The electron volt is defined as the amount of work done on one electron by a potential difference of one volt. Electrical charge multiplied by potential difference is equal to work or energy. The electrostatic unit of charge is the statcoulomb; one volt is equivalent to 1/299.7925 statvolt (See ELECTRICAL UNITS).

The relationship between the electron volt and the erg is calculated as follows:

$$\frac{4.80298 \times 10^{-10} \text{ statcoulomb}}{\text{electron}} \times \frac{\text{statvolt}}{299.7925 \text{ volts}}$$

$$= \frac{1.60210 \times 10^{-12} \text{ erg}}{\text{electron volt}}$$

The atomic mass unit is defined as the energy equivalent of one-twelfth the mass of one atom of the most abundant isotope of carbon, which is carbon-12. This unit had been previously defined as the energy equivalent of one-sixteenth the mass of one atom of oxygen-16. The carbon-12 standard was adopted by the International Union of Pure and Applied Physics in 1960 and by the International Union of Pure and Applied Chemistry in 1961. The oxygen standard led to confusion with physicists assigning the number 16 as the atomic mass of oxygen isotope 16, and chemists using 16 as the atomic mass of natural oxygen, which is a mixture of three isotopes but predominantly oxygen-16. The two standards differed by about 275 parts per million.

One gram-atom of carbon-12 has a mass equal to 12 grams, exactly, and contains "N" atoms, where "N" is the Avogadro constant. The energy equivalent of this mass is derived fron Einstein's equation relating energy, E, to mass, m, as follows, where "c" is the speed of light in a vacuum:

$$E = mc^2$$

The relationship between the atomic mass unit and the erg is calculated as follows:

$$E = \frac{1}{12} \times \frac{12 \text{ grams}}{\text{gram-atom}} \times \frac{\text{gram-atom}}{6.02252 \times 10^{23} \text{ atoms}}$$

$$\times (2.997925 \times 10^{10})^2 \frac{\text{centimeters}^2}{\text{second}^2}$$

$$E = 1.49232 \times 10^{-3} \frac{\text{gram centimeter}^2}{\text{second}^2} = \frac{1.49232 \times 10^{-3} \text{ erg}}{\text{atomic mass unit}}$$

The energy equivalents of the gram mass and pound mass are also derived from Einstein's equation, as follows:

$$E = 1 \text{ gram} \times (2.997925 \times 10^{10})^2 \frac{\text{centimeters}^2}{\text{second}^2}$$

$$= \frac{8.987554 \times 10^{20} \text{ ergs}}{\text{gram mass}}$$

$$E = 1 \text{ pound} \times \frac{453.59237 \text{ grams}}{\text{pound}} \times (2.997925 \times 10^{10})^2 \frac{\text{centimeters}^2}{\text{second}^2}$$

$$= \frac{4.076686 \times 10^{23} \text{ ergs}}{\text{pound mass}}$$

The basic atomic energy units are each related to the erg by the calculations above. The relationships between the erg and other energy units are given in the energy tables. Therefore, the atomic energy units can be related to other types of energy units and several such types are included in the atomic energy tables.

TABLE 3-1

UNITS OF ATOMIC ENERGY	Electron Volt	Kilo-Electron Volt	Million Electron Volt	Atomic Mass Unit	Billion Electron Volt	Erg	Foot Pound	Gram Calorie
Electron Volt	1	0.001	0.000001	1.07356×10^{-9}	1×10^{-9}	1.60210×10^{-12}	1.18165×10^{-19}	3.82911×10^{-20}
Kiloelectron Volt	1000	1	0.001	1.07356×10^{-6}	0.000001	1.60210×10^{-9}	1.18165×10^{-16}	3.82911×10^{-17}
Million Electron Volt	1,000,000	1000	1	1.07356×10^{-3}	0.001	1.60210×10^{-6}	1.18165×10^{-13}	3.82911×10^{-14}
Atomic Mass Unit	9.31479×10^{8}	9.31479×10^{5}	931.479	1	0.931479	1.49232×10^{-3}	1.10068×10^{-10}	3.56674×10^{-11}
Billion Electron Volt	1×10^{9}	1,000,000	1000	1.07356	1	1.60210×10^{-3}	1.18165×10^{-10}	3.82911×10^{-11}
Erg	6.24180×10^{11}	6.24180×10^{8}	6.24180×10^{5}	670.096	624.180	1	7.37562×10^{-8}	2.390057×10^{-8}
Foot Pound	8.46275×10^{18}	8.46275×10^{15}	8.46275×10^{12}	9.08528×10^{9}	8.46275×10^{9}	1.355818×10^{7}	1	0.3240483
Gram Calorie	2.61157×10^{19}	2.61157×10^{16}	2.61157×10^{13}	2.80368×10^{10}	2.61157×10^{10}	4.1840×10^{7}	3.085960	1

TABLE 3-2

UNITS OF ATOMIC ENERGY	Erg	Foot Poundal	Foot Pound	Kilogram Calorie	Horsepower Hour	Absolute Kilowatt Hour	Gram Mass Energy Equivalent	Pound Mass Energy Equivalent
Erg	1	2.373036×10^{-6}	7.37562×10^{-8}	2.390057×10^{-11}	3.725061×10^{-14}	2.777778×10^{-14}	1.112650×10^{-21}	2.452973×10^{-24}
Foot Poundal	4.214011×10^{5}	1	3.108095×10^{-2}	1.007173×10^{-5}	1.569745×10^{-8}	1.170559×10^{-8}	4.688718×10^{-16}	1.033685×10^{-18}
Foot Pound	1.355818×10^{7}	32.17405	1	3.240483×10^{-4}	5.050505×10^{-7}	3.766161×10^{-7}	1.508550×10^{-14}	3.325785×10^{-17}
Kilogram Calorie	4.1840×10^{10}	9.928783×10^{4}	3085.960	1	1.558566×10^{-3}	1.162222×10^{-3}	4.655327×10^{-11}	1.026324×10^{-13}
Horsepower Hour	2.684520×10^{13}	6.37046×10^{7}	1.98×10^{6}	641.616	1	0.745700	2.986930×10^{-8}	6.585053×10^{-11}
Absolute Kilowatt Hour	3.6×10^{13}	8.542930×10^{7}	2.655224×10^{6}	860.4207	1.341022	1	4.005539×10^{-8}	8.830702×10^{-11}
Gram Mass Energy Equivalent	8.987554×10^{20}	2.132779×10^{15}	6.62888×10^{13}	2.148077×10^{10}	3.347919×10^{7}	2.496543×10^{7}	1	2.2046226×10^{-3}
Pound Mass Energy Equivalent	4.076686×10^{23}	9.674123×10^{17}	3.006809×10^{16}	9.743514×10^{12}	1.518591×10^{10}	1.132413×10^{10}	453.59237	1

4. DENSITY
AND CONCENTRATION

Units of density and concentration listed in the tables are:

INTERNATIONAL MASS PER INTERNATIONAL VOLUME

1 grain per cubic foot
1 pound per cubic yard = 1/27 pound per cubic foot
1 ounce per cubic foot = 437.5 grains per cubic foot
1 pound per cubic foot = 7000 grains per cubic foot
1 short ton per cubic yard = 2000 pounds per cubic yard
1 long ton per cubic yard = 2240 pounds per cubic yard
1 ounce per cubic inch = 1728 ounces per cubic foot
1 pound per cubic inch = 16 ounces per cubic inch

METRIC MASS PER METRIC VOLUME

1 gram per cubic meter
1 milligram per cubic centimeter = 100 grams per cubic meter
1 gram per cubic centimeter = 1000 milligrams per cubic centimeter

INTERNATIONAL MASS PER U.S. LIQUID MEASURE

1 ounce per gallon
1 pound per gallon = 16 ounces per gallon

METRIC MASS PER METRIC CAPACITY

1 milligram per liter
1 milligram per milliliter = 1000 milligrams per liter
1 gram per milliliter = 1000 milligrams per milliliter

The various systems are related by the following:

1 pound = 453.59237 grams
1 international cubic inch = 16.387064 cubic centimeters
1 U.S. gallon = 231 international cubic inches
1 liter = 1000 cubic centimeters

The pound and ounce units of mass listed above are the international avoirdupois pound and ounce. The international units of mass were adopted by the National Bureau of Standards on July 1, 1959.

The gallon listed above is the United States gallon. The international units of volume are based on the international units of length adopted by the National Bureau of Standards, effective July 1, 1959. The United States gallon was originally defined as being equal to 231 United States customary cubic inches but it is now defined as being equal to 231 interntional cubic inches.

The liter is defined as equal to one cubic decimeter. Thus the liter is equal to 1000 cubic centimeters and the milliliter equals one cubic centimeter. This is a change from the previous definition of the liter as a unit of capacity equal to the volume occupied by one kilogram of pure water at its maximum density (at a temperature of $4°C$, practically) and under the standard atmospheric pressure of 760 millimeters of mercury. According to the old definition, one liter equalled 1000.028 cubic centimeters.

The international and old United States customary units of density and concentration are related as follows:

1 international unit = 1.00000587 U.S. customary unit
1 U.S. customary unit = 0.99999413 international unit

ADDITIONAL UNITS OF DENSITY AND CONCENTRATION

INTERNATIONAL

grain per gallon = 1.428571×10^{-4} pound per gallon
ounce per petroleum barrel = 0.02380952 ounce per gallon
pound per petroleum barrel = 0.02380952 pound per gallon
slug per cubic foot = 32.17405 pounds per cubic foot

METRIC

microgram per kiloliter = 0.000001 milligram per liter
microgram per cubic meter = 0.000001 gram per cubic meter
microgram per liter = 0.001 milligram per liter
milligram per kiloliter = 0.001 milligram per liter
milligram per cubic meter = 0.001 gram per cubic meter
microgram per milliliter = 1 milligram per liter
gram per kiloliter = 1 milligram per liter
microgram per cubic centimeter = 1 gram per cubic meter
gram per liter = 1 milligram per milliliter
kilogram per kiloliter = 1 milligram per milliliter
kilogram per cubic meter = 1 milligram per cubic centimeter

TABLE 4-1

UNITS OF DENSITY AND CONCENTRATION	Milligram per Liter	Gram per Cubic Meter	Grain per Cubic Foot	Pound per Cubic Yard	Milligram per Milliliter	Milligram per Cubic Centimeter	Ounce per Cubic Foot	Ounce per Gallon
Milligram per Liter	1	1	0.4369957	1.685555×10^{-3}	0.001	0.001	9.988474×10^{-4}	1.335265×10^{-4}
Gram per Cubic Meter	1	1	0.4369957	1.685555×10^{-3}	0.001	0.001	9.988474×10^{-4}	1.335265×10^{-4}
Grain per Cubic Foot	2.288352	2.288352	1	3.857143×10^{-3}	2.288352×10^{-3}	2.288352×10^{-3}	2.285714×10^{-3}	3.055556×10^{-4}
Pound per Cubic Yard	593.2764	593.2764	259.2593	1	0.5932764	0.5932764	0.5925926	7.921811×10^{-2}
Milligram per Milliliter	1000	1000	436.9957	1.685555	1	1	0.9988474	0.1335265
Milligram per Cubic Centimeter	1000	1000	436.9957	1.685555	1	1	0.9988474	0.1335265
Ounce per Cubic Foot	1001.154	1001.154	437.5	1.6875	1.001154	1.001154	1	0.1336806
Ounce per Gallon	7489.152	7489.152	3272.727	12.62338	7.489152	7.489152	7.480519	1

TABLE 4-2

UNITS OF DENSITY AND CONCENTRATION	Milligram per Liter	Gram per Cubic Meter	Grain per Cubic Foot	Pound per Cubic Yard	Milligram per Milliliter	Milligram per Cubic Centimeter	Ounce per Cubic Foot	Ounce per Gallon
Pound per Cubic Foot	16,018.46	16,018.46	7000	27	16.01846	16.01846	16	2.138889
Pound per Gallon	119,826.4	119,826.4	52,363.64	201.9740	119.8264	119.8264	119.6883	16
Gram per Milliliter	1,000,000	1,000,000	436,995.7	1685.555	1000	1000	998.8474	133.5265
Gram per Cubic Centimeter	1,000,000	1,000,000	436,995.7	1685.555	1000	1000	998.8474	133.5265
Short Ton per Cubic Yard	1.186553×10^6	1.186553×10^6	518,518.5	2000	1186.553	1186.553	1185.185	158.4362
Long Ton per Cubic Yard	1.328939×10^6	1.328939×10^6	580,740.7	2240	1328.939	1328.939	1327.407	177.4486
Ounce per Cubic Inch	1.729994×10^6	1.729994×10^6	756,000	2916	1729.994	1729.994	1728	231
Pound per Cubic Inch	2.767990×10^7	2.767990×10^7	12,096,000	46,656	27,679.90	27,679.90	27,648	3696

TABLE 4-3

UNITS OF DENSITY AND CONCENTRATION	Pound per Cubic Foot	Pound per Gallon	Gram per Milliliter	Gram per Cubic Centimeter	Short Ton per Cubic Yard	Long Ton per Cubic Yard	Ounce per Cubic Inch	Pound per Cubic Inch
Milligram per Liter	6.242796×10^{-5}	8.345404×10^{-6}	0.000001	0.000001	8.427775×10^{-7}	7.524799×10^{-7}	5.780367×10^{-7}	3.612729×10^{-8}
Gram per Cubic Meter	6.242796×10^{-5}	8.345404×10^{-6}	0.000001	0.000001	8.427775×10^{-7}	7.524799×10^{-7}	5.780367×10^{-7}	3.612729×10^{-8}
Grain per Cubic Foot	1.428571×10^{-4}	1.909722×10^{-5}	2.288352×10^{-6}	2.288352×10^{-6}	1.928571×10^{-6}	1.721939×10^{-6}	1.322751×10^{-6}	8.267196×10^{-8}
Pound per Cubic Yard	3.703704×10^{-2}	4.951132×10^{-3}	5.932764×10^{-4}	5.932764×10^{-4}	0.0005	4.464286×10^{-4}	3.429355×10^{-4}	2.143347×10^{-5}
Milligram per Milliliter	6.242796×10^{-2}	8.345404×10^{-3}	0.001	0.001	8.427775×10^{-4}	7.524799×10^{-4}	5.780367×10^{-4}	3.612729×10^{-5}
Milligram per Cubic Centimeter	6.242796×10^{-2}	8.345404×10^{-3}	0.001	0.001	8.427775×10^{-4}	7.524799×10^{-4}	5.780367×10^{-4}	3.612729×10^{-5}
Ounce per Cubic Foot	0.0625	8.355035×10^{-3}	1.001154×10^{-3}	1.001154×10^{-3}	8.4375×10^{-4}	7.533482×10^{-4}	5.787037×10^{-4}	3.616898×10^{-5}
Ounce per Gallon	0.4675325	0.0625	7.489152×10^{-3}	7.489152×10^{-3}	6.311688×10^{-3}	5.635436×10^{-3}	4.329004×10^{-3}	2.705628×10^{-4}

TABLE 4-4

UNITS OF DENSITY AND CONCENTRATION	Pound per Cubic Foot	Pound per Gallon	Gram per Milliliter	Gram per Cubic Centimeter	Short Ton per Cubic Yard	Long Ton per Cubic Yard	Ounce per Cubic Inch	Pound per Cubic Inch
Pound per Cubic Foot	1	0.1336806	1.601846×10^{-2}	1.601846×10^{-2}	0.0135	1.205357×10^{-2}	9.259259×10^{-3}	5.787037×10^{-4}
Pound per Gallon	7.480519	1	0.1198264	0.1198264	0.1009870	9.016698×10^{-2}	6.926407×10^{-2}	4.329004×10^{-3}
Gram per Milliliter	62.42796	8.345404	1	1	0.8427775	0.7524799	0.5780367	3.612729×10^{-2}
Gram per Cubic Centimeter	62.42796	8.345404	1	1	0.8427775	0.7524799	0.5780367	3.612729×10^{-2}
Short Ton per Cubic Yard	74.07407	9.902263	1.186553	1.186553	1	0.8928571	0.6858711	4.286694×10^{-2}
Long Ton per Cubic Yard	82.96296	11.09053	1.328939	1.328939	1.12	1	0.7681756	4.801097×10^{-2}
Ounce per Cubic Inch	108	14.4375	1.729994	1.729994	1.458	1.301786	1	0.0625
Pound per Cubic Inch	1728	231	27.67990	27.67990	23.328	20.82857	16	1

5. ELECTRICAL UNITS

There are seven tables of electrical units with each table listing units of each of the three fundamental systems: the practical system in absolute units, the electromagnetic system in which units have the prefix "ab-", and the electrostatic system in which units have the prefix "stat-". All three systems are based on metric units.

Electrical units listed in the tables are:

CHARGE (OR QUANTITY)

1 statcoulomb $= 3.335640 \times 10^{-10}$ coulomb
1 coulomb
1 abcoulomb $= 10$ coulombs

CURRENT

1 statampere $= 3.335640 \times 10^{-10}$ ampere
1 ampere
1 abampere $= 10$ amperes

RESISTANCE

1 abohm $= 1 \times 10^{-9}$ ohm
1 ohm
1 statohm $= 8.987554 \times 10^{11}$ ohms

POTENTIAL

1 abvolt $= 1 \times 10^{-8}$ volt
1 volt
1 statvolt $= 299.7925$ volts

CAPACITANCE

1 statfarad $= 1.112650 \times 10^{-12}$ farad
1 farad
1 abfarad $= 1 \times 10^{9}$ farads

INDUCTANCE

1 abhenry $= 1 \times 10^{-9}$ henry
1 henry
1 stathenry $= 8.987554 \times 10^{11}$ henrys

CONDUCTANCE

1 statmho $= 1.112650 \times 10^{-12}$ mho
1 mho
1 abmho $= 1 \times 10^9$ mhos

The ampere is that constant current which, if maintained in two straight parallel conductors of infinite length, of negligible circular cross-section, and placed one meter apart in a vacuum, would produce between these conductors a force equal to 2×10^{-7} newton per meter of length (2). The practical unit of charge or quantity, the absolute coulomb, is the quantity of electricity passing any part of an electric circuit in one second when the current is one absolute ampere, which is defined above as the ampere. The ohm as the unit of resistance may be defined on the basis that a current of one absolute ampere through a resistance of one absolute ohm will liberate energy at the rate of one joule per second, or one watt. By Ohm's Law, a current of one absolute ampere through a resistance of one absolute ohm yields a potential difference of one absolute volt.

The absolute farad is defined as that capacitance the potential of which is raised one volt by the addition of one coulomb of charge, that is, one absolute farad equals one absolute coulomb per absolute volt. An electric circuit has an inductance of one absolute henry when a current change of one absolute ampere per second induces an electromotive force, or potential difference, of one absolute volt. Conductance is the reciprocal of resistance and the unit of conductance, the absolute mho, is the reciprocal of the absolute ohm.

Therefore, the seven electrical units in the practical system are related to the metric system through the meter, the newton, which equals one kilogram mass meter per second per second, and the joule, which equals one newton meter.

The electromagnetic unit (e.m.u.) of current is the abampere, defined as that current which, flowing in one centimeter of a circuit bent into an arc of one centimeter radius, produces a magnetic field of unit intensity at the center of the circle. The abcoulomb is equal to one abampere second. One abampere through a resistance of one abohm will liberate energy at the rate of one erg per second.

Ohm's Law is used to define the abvolt as the potential difference produced by a current of one abampere through a resistance of one abohm. One abfarad equals one abcoulomb per abvolt. A current change of one abampere per second induces an electromotive force of one abvolt when the inductance is one abhenry. The abmho is the reciprocal of the abohm.

The units of the electromagnetic system are related to the metric system of measurement through the centimeter and the erg, which equals one dyne centimeter. Conversion from the electromagnetic units to practical units is based on the relationship that one abampere is equal to ten amperes.

The electrostatic unit (e.s.u.) of charge, the statcoulomb, is defined as that charge which is repelled with the force of one dyne by an equal charge one centimeter distant when both are in a vacuum. The statampere equals one statcoulomb per second. One erg per second will be liberated when one statampere flows through a resistance of one statohm, and the potential difference produced will be one statvolt.

The statfarad is equal to one statcoulomb per statvolt. The inductance will be one stathenry when a current change of one statampere per second induces an electromotive force of one statvolt. The statmho is the reciprocal of the statohm.

The units of the electrostatic system are based on the following metric units: the centimeter, the dyne, which equals one gram mass centimeter per second per second, and the erg. One abampere is equal to "c" statamperes, where "c" is the speed of light in a vacuum, 2.997925×10^{10} centimeters per second (1).

ADDITIONAL ELECTRICAL UNITS

CURRENT

microampere = 0.000001 ampere
milliampere = 0.001 ampere
international ampere = 0.999835 ampere

CHARGE

international coulomb = 0.999835 coulomb
ampere hour = 3600 coulombs

RESISTANCE

microhm = 0.000001 ohm
international ohm = 1.000495 ohms
megohm = 1,000,000 ohms

POTENTIAL

microvolt = 0.000001 volt
millivolt = 0.001 volt
international volt = 1.00033 volt
kilovolt = 1000 volts

CAPACITANCE

micromicrofarad = 1×10^{-12} farad
microfarad = 0.000001 farad
international farad = 0.999505 farad

INDUCTANCE

millihenry = 0.001 henry
international henry = 1.000495 henrys

CONDUCTANCE

international mho = 0.999505 mho
megmho = 1,000,000 mhos

International units listed above are taken directly from *National Bureau of Standards Circular C459*, "Announcement of Changes in Electrical and Photometric Units," May 15, 1947. Only the international mho was derived, using the international ohm as the basis for the derivation. The following statement is made in the Circular:

> "This Circular gives a short account of the development of new international agreements on practical units of electricity and of light. In pursuance of these agreements, the electrical units based upon the resistance of a column of mercury and the rate of deposition of silver in a voltameter will be superseded on January 1, 1948, by units derived from the fundamental mechanical units of length, mass, and time."

Therefore, international units have been replaced by the practical or absolute system of electrical units, which are based on fundamental mechanical units. For this reason the international units are not included in the tables, nor is the term "absolute" used to preface any of the practical units.

There are many other types of electrical measurements and some of these are listed as follows, with representative units given for each:

RESISTIVITY - ohm centimeter
CONDUCTIVITY - mho per centimeter
INDUCTIVE REACTANCE - ohm
CAPACITIVE REACTANCE - ohm
IMPEDANCE - ohm
SUSCEPTANCE - mho
ADMITTANCE - mho
MASS RESISTIVITY - ohm centimeter gram
MASS CONDUCTIVITY - mho per centimeter gram
ELECTRICAL FIELD STRENGTH - volt per centimeter

TABLE 5-1

UNITS OF ELECTRICITY CHARGE	Statcoulomb	Coulomb	Abcoulomb
Statcoulomb	1	3.335640×10^{-10}	3.335640×10^{-11}
Coulomb	2.997925×10^{9}	1	0.1
Abcoulomb	2.997925×10^{10}	10	1

TABLE 5-2

UNITS OF ELECTRICITY CURRENT	Statampere	Ampere	Abampere
Statampere	1	3.335640×10^{-10}	3.335640×10^{-11}
Ampere	2.997925×10^{9}	1	0.1
Abampere	2.997925×10^{10}	10	1

TABLE 5-3

UNITS OF ELECTRICITY RESISTANCE	Abohm	Ohm	Statohm
Abohm	1	1×10^{-9}	1.112650×10^{-21}
Ohm	1×10^{9}	1	1.112650×10^{-12}
Statohm	8.987554×10^{20}	8.987554×10^{11}	1

TABLE 5-4

UNITS OF ELECTRICITY POTENTIAL	Abvolt	Volt	Statvolt
Abvolt	1	1×10^{-8}	3.335640×10^{-11}
Volt	1×10^{8}	1	3.335640×10^{-3}
Statvolt	2.997925×10^{10}	299.7925	1

TABLE 5-6

UNITS OF ELECTRICITY INDUCTANCE	Abhenry	Henry	Stathenry
Abhenry	1	1×10^{-9}	1.112650×10^{-21}
Henry	1×10^{9}	1	1.112650×10^{-12}
Stathenry	8.987554×10^{20}	8.987554×10^{11}	1

TABLE 5-5

UNITS OF ELECTRICITY CAPACITANCE	Statfarad	Farad	Abfarad
Statfarad	1	1.112650×10^{-12}	1.112650×10^{-21}
Farad	8.987554×10^{11}	1	1×10^{-9}
Abfarad	8.987554×10^{20}	1×10^{9}	1

TABLE 5-7

UNITS OF ELECTRICITY CONDUCTANCE	Statmho	Mho	Abmho
Statmho	1	1.112650×10^{-12}	1.112650×10^{-21}
Mho	8.987554×10^{11}	1	1×10^{-9}
Abmho	8.987554×10^{20}	1×10^{9}	1

6. ENERGY

Units of energy listed in the tables are:

INTERNATIONAL FORCE TIMES INTERNATIONAL LENGTH

1 foot poundal
1 foot pound = 980.665/30.48 foot poundals
1 horsepower hour = 1,980,000 foot pounds

METRIC AND RELATED UNITS (ABSOLUTE)

1 erg
1 gram centimeter = 980.665 ergs
1 absolute joule = 10,000,000 ergs
1 gram calorie = 4.1840 absolute joules
1 kilogram meter = 100,000 gram centimeters
1 BTU (British Thermal Unit) = 251.9958 gram calories
1 absolute watt hour = 3600 absolute joules
1 kilogram calorie = 1000 gram calories
1 metric horsepower hour = 270,000 kilogram meters
1 electrical horsepower hour = 746 absolute watt hours
1 absolute kilowatt hour = 1000 absolute watt hours

METRIC AND RELATED UNITS (INTERNATIONAL)

1 international joule = 1.000165 absolute joules
1 I.T. gram calorie = 4.1868 absolute joules
1 I.T. BTU = 251.9958 I.T. gram calories
1 international watt hour = 3600 international joules
1 I.T. kilogram calorie = 1000 I.T. gram calories
1 international kilowatt hour = 1000 international watt hours

The various systems are related by the following:

1 foot = 30.48 centimeters
1 pound = 453.59237 grams

The foot listed above is the international foot and the pound is the international avoirdupois pound. International units in these cases are those adopted by the National Bureau of Standards, effective July 1, 1959.

Units of force are those derived from units of mass in the cases of the pound, gram and kilogram through the relationship that force is equal to mass times the acceleration of gravity.   The standard acceleration of gravity equals 980.665 centimeters per second per second, which is equivalent to 32.17405 feet per second per second based on 30.48 centimeters per international foot.   The poundal is equal to one international avoirdupois pound mass foot per second per second.

The erg is equal to one dyne centimeter and the dyne equals one gram mass centimeter per second per second.   Therefore, the gram (force) centimeter is equivalent to 980.665 dyne centimeters, or 980.665 ergs.   The absolute joule is defined as being equal to 10,000,000 ergs.   The absolute watt, a unit of power (energy per unit time), is equal to one absolute joule per second; therefore the absolute watt hour equals 3600 absolute joules.

As originally defined, the gram calorie was the amount of heat (a form of energy) required to raise the temperature of one gram of water through one degree Centigrade.   As thermal measurements were made with increased precision, this definition was not sufficiently accurate.   Confusion developed as many different kinds of calorie were used.   For instance, the amount of heat required to raise one gram of water from 15°C to 16°C is not the same as the amount of heat required to raise one gram of water from 60°C to 61°C.   After 1930, an artificial calorie was defined as being equal to 4.1833 international joules.   In 1948, the calorie was redefined as being equal to 4.1840 absolute joules and this definition is generally accepted in the United States.   The gram calorie is also known as the thermo-chemical calorie, and some references mention the kilogram calorie used in these tables as being the large calorie, or Calorie.

Another artificial calorie, used in engineering steam tables, is the International Steam Table calorie, or I.T. gram calorie, which is defined as being equal to 4.1868 absolute joules.   It was previously defined as the equivalent of 1/860 international watt hour. (See POWER).

The British Thermal Unit, BTU, was originally defined as the amount of heat required to raise the temperature of one pound of water through one degree Fahrenheit.   Although the gram calorie is no longer defined on the gram of water through one degree Centigrade basis, the relationship between the British Thermal Unit and the gram calorie (and the I.T. BTU and the I.T. gram calorie) is still based on comparisons between the pound and gram and the degrees Fahrenheit and Centigrade, as follows:

$$\frac{(1 \text{ pound})(1\,^\circ\text{F})}{\text{BTU}} \times \frac{453.59237 \text{ grams}}{\text{pound}} \times \frac{1\,^\circ\text{C}}{1.8\,^\circ\text{F}} \times \frac{\text{gram calorie}}{(1 \text{ gram})(1\,^\circ\text{C})}$$

$$= 251.9958 \text{ gram calories per BTU (not exactly)}$$

Therefore, one British Thermal Unit is equal to 251.9958 gram calories, where the gram calorie equals 4.1840 absolute joules, and one I.T. British Thermal Unit is equal to 251.9958 I.T. gram calories, where the I.T. gram calorie equals 4.1868 absolute joules. It is because of these relationships that the British Thermal Unit and the I.T. British Thermal Unit are grouped with the metric units in this book.

The following statement is made in the National Bureau of Standards Circular C459, "Announcement of Changes in Electrical and Photometric Units," May 15, 1947:

"This Circular gives a short account of the development of new international agreements on practical units of electricity and of light. In pursuance of these agreements, the electrical units based upon the resistance of a column of mercury and the rate of deposition of silver in a voltameter will be superseded on January 1, 1948, by units derived from the fundamental mechanical units of length, mass, and time."

In this circular, the international watt is listed as being equal to 1.000165 absolute watts and the international joule equals 1.000165 absolute joules. Therefore, with the exception of the I.T. calories, absolute units have replaced the international units. However, the international joule, watt hour, and kilowatt hour are listed in this book for historical purposes. These units are related to the metric system of units and are not to be confused with the international system adopted by the National Bureau of Standards, effective July 1, 1959.

The horsepower, a unit of power, was originally defined as being equal to 550 United States customary foot pounds per second. In these tables it is defined as being equal to 550 international foot pounds per second. Based on this definition, the horsepower hour is equivalent to 1,980,000 international foot pounds.

The metric horsepower is defined as being equal to 75 kilogram meters per second; therefore, one metric horsepower hour equals 270,000 kilogram meters. The electrical horsepower equals 746 absolute watts and one electrical horsepower hour is equivalent to 746 absolute watt hours.

In cases of the foot pound, foot poundal and horsepower hour, the international and old United States customary units are related as follows:

$$1 \text{ international unit} = 0.9999959 \text{ U.S. customary unit}$$
$$1 \text{ U.S. customary unit} = 1.0000041 \text{ international units}$$

The British Thermal Unit based on the international pound is 0.99999987 BTU based on the United States customary pound.

ADDITIONAL UNITS OF ENERGY

INTERNATIONAL

$$\text{cubic foot atmosphere} = 2116.217 \text{ foot pounds}$$

METRIC

$$\text{dyne centimeter} = 1 \text{ erg (in tables)}$$
$$\text{megalerg (megerg)} = 1,000,000 \text{ ergs}$$
$$\text{cubic centimeter atmosphere} = 0.1013250 \text{ absolute joule}$$
$$\text{newton meter} = 1 \text{ absolute joule (in tables)}$$
$$\text{liter atmosphere} = 101.3250 \text{ absolute joules}$$

There are gram calorie units defined on the basis that one gram calorie is equal to the amount of heat required to raise the temperature of one gram of pure water one degree Centigrade. Because the specific heat of water varies with temperature, these units are usually qualified as to their reference temperature, such as the $15°C$ gram calorie. One one-hundredth of the amount of heat required to raise one gram of pure water from $0°C$ to $100°C$ is the equivalent of the mean gram calorie.

In the tables the gram calorie is defined as equal to 4.1840 absolute joules, and the I.T. gram calorie is defined as equal to 4.1868 absolute joules. For comparison purposes, the following table lists the more common gram calories and their approximate conversion factors with respect to the absolute joule:

$$\text{gram calorie } (20°C) = 4.181 \text{ absolute joules}$$
$$\text{gram calorie } (15°C) = 4.1855 \text{ absolute joules}$$
$$\text{gram calorie (mean)} = 4.186 \text{ absolute joules}$$

A British Thermal Unit (BTU) at the same temperature as a gram calorie is equal to 251.9958 gram calories. Such a British Thermal Unit is equal to the amount of heat required to raise the temperature of one pound of pure water one degree Fahrenheit. The British Thermal Unit defined on this basis is usually qualified in degrees Fahrenheit. Two units more commonly employed are the BTU (39°F), which is related to the gram calorie at approximately 4°C, and the BTU (60°F), which is related to the gram calorie at about 15°C.

TABLE 6-1

UNITS OF ENERGY	Erg	Gram Centimeter	Foot Poundal	Absolute Joule	International Joule	Foot Pound	Gram Calorie	I. T. Gram Calorie
Erg	1	1.019716×10^{-3}	2.373036×10^{-6}	1×10^{-7}	9.99835×10^{-8}	7.37562×10^{-8}	2.390057×10^{-8}	2.388459×10^{-8}
Gram Centimeter	980.665	1	2.327153×10^{-3}	9.80665×10^{-5}	9.80503×10^{-5}	7.23301×10^{-5}	2.343846×10^{-5}	2.342278×10^{-5}
Foot Poundal	4.214011×10^{5}	429.7095	1	4.214011×10^{-2}	4.213316×10^{-2}	3.108095×10^{-2}	1.007173×10^{-2}	1.006499×10^{-2}
Absolute Joule	1×10^{7}	1.019716×10^{4}	23.73036	1	0.999835	0.737562	0.2390057	0.2388459
International Joule	1.000165×10^{7}	1.019884×10^{4}	23.73428	1.000165	1	0.737684	0.2390452	0.2388853
Foot Pound	1.355818×10^{7}	1.382550×10^{4}	32.17405	1.355818	1.355594	1	0.3240483	0.323316
Gram Calorie	4.1840×10^{7}	4.266493×10^{4}	99.2878	4.1840	4.183310	3.085960	1	0.999331
I. T. Gram Calorie	4.1868×10^{7}	4.269348×10^{4}	99.3543	4.1868	4.186109	3.088025	1.000669	1
Kilogram Meter	9.80665×10^{7}	1×10^{5}	232.7153	9.80665	9.80503	7.23301	2.343846	2.342278
BTU	1.054350×10^{10}	1.075138×10^{7}	2.502011×10^{4}	1054.350	1054.176	777.649	251.9958	251.8272
I.T. BTU	1.055056×10^{10}	1.075858×10^{7}	2.503686×10^{4}	1055.056	1054.882	778.169	252.1644	251.9958

TABLE 6-2

UNITS OF ENERGY	Erg	Gram Centimeter	Foot Poundal	Absolute Joule	International Joule	Foot Pound	Gram Calorie	I. T. Gram Calorie
Absolute Watt Hour	3.6×10^{10}	3.670978×10^7	8.54293×10^4	3600	3599.406	2655.224	860.421	859.845
International Watt Hour	3.600594×10^{10}	3.671584×10^7	8.54434×10^4	3600.594	3600	2655.662	860.563	859.987
Kilogram Calorie	4.1840×10^{10}	4.266493×10^7	9.92878×10^4	4.1840×10^3	4183.310	3085.960	1000	999.331
I. T. Kilogram Calorie	4.1868×10^{10}	4.269348×10^7	9.93543×10^4	4.1868×10^3	4186.109	3088.025	1000.669	1000
Metric Horsepower Hour	2.647796×10^{13}	2.7×10^{10}	6.28331×10^7	2.647796×10^6	2.647359×10^6	1.952914×10^6	6.32838×10^5	6.32415×10^5
Horsepower Hour	2.684520×10^{13}	2.737448×10^{10}	6.37046×10^7	2.684520×10^6	2.684077×10^6	1.98×10^6	6.41616×10^5	6.41186×10^5
Electrical Horsepower Hour	2.6856×10^{13}	2.738550×10^{10}	6.37303×10^7	2.6856×10^6	2.685157×10^6	1.980797×10^6	6.41874×10^5	6.41445×10^5
Absolute Kilowatt Hour	3.6×10^{13}	3.670978×10^{10}	8.54293×10^7	3,600,000	3.599406×10^6	2.655224×10^6	8.60421×10^5	8.59845×10^5
International Kilowatt Hour	3.600594×10^{13}	3.671584×10^{10}	8.54434×10^7	3.600594×10^6	3,600,000	2.655662×10^6	8.60563×10^5	8.59987×10^5

TABLE 6-3

UNITS OF ENERGY	Kilogram Meter	BTU	I. T. BTU	Absolute Watt Hour	International Watt Hour	Kilogram Calorie	I. T. Kilogram Calorie
Erg	1.019716×10^{-8}	9.48451×10^{-11}	9.47817×10^{-11}	2.777778×10^{-11}	2.777320×10^{-11}	2.390057×10^{-11}	2.388459×10^{-11}
Gram Centimeter	0.00001	9.30113×10^{-8}	9.29491×10^{-8}	2.724069×10^{-8}	2.723620×10^{-8}	2.343846×10^{-8}	2.342278×10^{-8}
Foot Poundal	4.297095×10^{-3}	3.996785×10^{-5}	3.994112×10^{-5}	1.170559×10^{-5}	1.170366×10^{-5}	1.007173×10^{-5}	1.006499×10^{-5}
Absolute Joule	0.1019716	9.48451×10^{-4}	9.47817×10^{-4}	2.777778×10^{-4}	2.777320×10^{-4}	2.390057×10^{-4}	2.388459×10^{-4}
International Joule	0.1019884	9.48608×10^{-4}	9.47974×10^{-4}	2.778236×10^{-4}	2.777778×10^{-4}	2.390452×10^{-4}	2.388853×10^{-4}
Foot Pound	0.1382550	1.285927×10^{-3}	1.285067×10^{-3}	3.766161×10^{-4}	3.765540×10^{-4}	3.240483×10^{-4}	3.238316×10^{-4}
Gram Calorie	0.4266493	3.968321×10^{-3}	3.965667×10^{-3}	1.162222×10^{-3}	1.162030×10^{-3}	0.001	9.99331×10^{-4}
I. T. Gram Calorie	0.4269348	3.970976×10^{-3}	3.968321×10^{-3}	1.163×10^{-3}	1.162808×10^{-3}	1.000669×10^{-3}	0.001
Kilogram Meter	1	9.30113×10^{-3}	9.29491×10^{-3}	2.724069×10^{-3}	2.723620×10^{-3}	2.343846×10^{-3}	2.342278×10^{-3}
BTU	107.5138	1	0.999331	0.2928751	0.2928268	0.2519958	0.2518272
I. T. BTU	107.5858	1.000669	1	0.2930711	0.2930227	0.2521644	0.2519958

TABLE 6-4

UNITS OF ENERGY	Kilogram Meter	BTU	I.T. BTU	Absolute Watt Hour	International Watt Hour	Kilogram Calorie	I.T. Kilogram Calorie
Absolute Watt Hour	367.0978	3.414425	3.412142	1	0.999835	0.860421	0.859845
International Watt Hour	367.1584	3.414988	3.412705	1.000165	1	0.860563	0.859987
Kilogram Calorie	426.6493	3.968321	3.965667	1.162222	1.162030	1	0.999331
I.T. Kilogram Calorie	426.9348	3.970976	3.968321	1.163	1.162808	1.000669	1
Metric Horsepower Hour	2.7×10^5	2511.305	2509.626	735.499	735.377	632.838	632.415
Horsepower Hour	2.737448×10^5	2546.136	2544.434	745.700	745.577	641.616	641.186
Electrical Horsepower Hour	2.738550×10^5	2547.161	2545.458	746	745.877	641.874	641.445
Absolute Kilowatt Hour	3.670978×10^5	3414.425	3412.142	1000	999.835	860.421	859.845
International Kilowatt Hour	3.671584×10^5	3414.988	3412.705	1000.165	1000	860.563	859.987

TABLE 6-5

UNITS OF ENERGY	Metric Horsepower Hour	Horsepower Hour	Electrical Horsepower Hour	Absolute Kilowatt Hour	International Kilowatt Hour
Erg	3.776727×10^{-14}	3.725061×10^{-14}	3.723563×10^{-14}	2.777778×10^{-14}	2.777320×10^{-14}
Gram Centimeter	3.703704×10^{-11}	3.653037×10^{-11}	3.651568×10^{-11}	2.724069×10^{-11}	2.723620×10^{-11}
Foot Poundal	1.591517×10^{-8}	1.569745×10^{-8}	1.569113×10^{-8}	1.170559×10^{-8}	1.170366×10^{-8}
Absolute Joule	3.766727×10^{-7}	3.725061×10^{-7}	3.723563×10^{-7}	2.777778×10^{-7}	2.777320×10^{-7}
International Joule	3.777350×10^{-7}	3.725676×10^{-7}	3.724177×10^{-7}	2.778236×10^{-7}	2.777778×10^{-7}
Foot Pound	5.12055×10^{-7}	5.05051×10^{-7}	5.04847×10^{-7}	3.766161×10^{-7}	3.765540×10^{-7}
Gram Calorie	1.580182×10^{-6}	1.558566×10^{-6}	1.557939×10^{-6}	1.162222×10^{-6}	1.162030×10^{-6}
I.T. Gram Calorie	1.581240×10^{-6}	1.559609×10^{-6}	1.558981×10^{-6}	1.163×10^{-6}	1.162808×10^{-6}
Kilogram Meter	3.703704×10^{-6}	3.653037×10^{-6}	3.651568×10^{-6}	2.724069×10^{-6}	2.723620×10^{-6}
BTU	3.981993×10^{-4}	3.927519×10^{-4}	3.925939×10^{-4}	2.928751×10^{-4}	2.928268×10^{-4}
I.T. BTU	3.984658×10^{-4}	3.930148×10^{-4}	3.928567×10^{-4}	2.930711×10^{-4}	2.930227×10^{-4}

TABLE 6-6

UNITS OF ENERGY	Metric Horsepower Hour	Horsepower Hour	Electrical Horsepower Hour	Absolute Kilowatt Hour	International Kilowatt Hour
Absolute Watt Hour	1.359622×10^{-3}	1.341022×10^{-3}	1.340483×10^{-3}	0.001	9.99835×10^{-4}
International Watt Hour	1.359846×10^{-3}	1.341243×10^{-3}	1.340704×10^{-3}	1.000165×10^{-3}	0.001
Kilogram Calorie	1.580182×10^{-3}	1.558566×10^{-3}	1.557939×10^{-3}	1.162222×10^{-3}	1.162030×10^{-3}
I.T. Kilogram Calorie	1.581240×10^{-3}	1.559609×10^{-3}	1.558981×10^{-3}	1.163×10^{-3}	1.162808×10^{-3}
Metric Horsepower Hour	1	0.986320	0.985923	0.735499	0.735377
Horsepower Hour	1.013870	1	0.999598	0.745700	0.745577
Electrical Horsepower Hour	1.014278	1.000402	1	0.746	0.745877
Absolute Kilowatt Hour	1.359622	1.341022	1.340483	1	0.999835
International Kilowatt Hour	1.359846	1.341243	1.340704	1.000165	1

7. FLOW

Units of flow listed in the table are:

INTERNATIONAL VOLUME PER TIME

1 cubic foot per minute
1 cubic yard per minute = 27 cubic feet per minute
1 cubic foot per second = 60 cubic feet per minute

METRIC VOLUME PER TIME

1 cubic centimeter per second

U.S. LIQUID MEASURE PER TIME

1 petroleum barrel per hour = 0.7 gallon per minute
1 gallon per minute
1 gallon per second = 60 gallons per minute

METRIC CAPACITY PER TIME

1 liter per minute
1 liter per second = 60 liters per minute

The various systems are related by the following:

1 international cubic foot per
                    second = 1728 international cubic inches
                             per second
1 international cubic inch per
                    second = 16.387064 cubic centimeters per
                             second
    1 U.S. gallon per second = 231 international cubic inches per
                             second
        1 liter per second = 1000 cubic centimeters per second

The international units of volume are based on the international units of length adopted by the National Bureau of Standards, effective July 1, 1959. Units of United States liquid measure were originally defined in terms of the old United States customary units but are now related to the international system. The petroleum barrel is defined as equal to 42 United States gallons.

The liter is defined as equal to one cubic decimeter. Thus the liter is equal to 1000 cubic centimeters. This is a change from the previous definition of the liter as a unit of capacity equal to the volume occupied by one kilogram of pure water at its maximum density (at a temperature of 4°C, practically) and under the standard atmospheric pressure of 760 millimeters of mercury. According to the old definition, one liter equalled 1000.028 cubic centimeters.

The international and old United States customary units of flow are related as follows:

1 international unit = 0.999994 U.S. customary unit
1 U.S. customary unit = 1.000006 international unit

ADDITIONAL UNITS OF FLOW

INTERNATIONAL

gallon per hour = 0.01666667 gallon per minute
petroleum barrel per minute = 60 petroleum barrels per hour
acre foot per day = 30.25 cubic feet per minute
petroleum barrel per second = 3600 petroleum barrels per hour

METRIC

cubic meter per hour = 277.7778 cubic centimeters per second
kiloliter per hour = 16.66667 liters per minute
cubic meter per minute = 16,666.67 cubic centimeters per second
kiloliter per minute = 1000 liters per minute
cubic meter per second = 1,000,000 cubic centimeters per second
kiloliter per second = 1000 liters per second

TABLE 7-1

UNITS OF FLOW	Cubic Centimeter per Second	Liter per Minute	Petroleum Barrel per Hour	Gallon per Minute	Cubic Foot per Minute	Liter per Second	Gallon per Second	Cubic Yard per Minute	Cubic Foot per Second
Cubic Centimeter per Second	1	0.06	2.264332×10^{-2}	1.585032×10^{-2}	2.118880×10^{-3}	0.001	2.641721×10^{-4}	7.847704×10^{-5}	3.531467×10^{-5}
Liter per Minute	16.66667	1	0.3773886	0.2641721	3.531467×10^{-2}	1.666667×10^{-2}	4.402868×10^{-3}	$1.307951 \; 10^{-3}$	5.885778×10^{-4}
Petroleum Barrel per Hour	44.16314	2.649788	1	0.7	9.357639×10^{-2}	4.416314×10^{-2}	1.166667×10^{-2}	3.465792×10^{-3}	1.559606×10^{-3}
Gallon per Minute	63.09020	3.875412	1.428571	1	0.1336806	6.309020×10^{-2}	1.666667×10^{-2}	4.951132×10^{-3}	2.228009×10^{-3}
Cubic Foot per Minute	471.9474	28.31685	10.68646	7.480519	1	0.4719474	0.1246753	3.703704×10^{-2}	1.666667×10^{-2}
Liter per Second	1000	60	22.64332	15.85032	2.118880	1	0.2641721	7.847704×10^{-2}	3.531467×10^{-2}
Gallon per Second	3785.412	227.1247	85.71429	60	8.020833	3.785412	1	0.2970679	0.1336806
Cubic Yard per Minute	12,742.58	764.5549	288.5343	201.9740	27	12.74258	3.366234	1	0.45
Cubic Foot per Second	28,316.85	1699.011	641.1874	448.8312	60	28.31685	7.480519	2.222222	1

8. FORCE

Units of force listed in the table are:

INTERNATIONAL

1 poundal
1 pound = 32.1740 poundals
1 short ton = 2000 pounds

METRIC

1 dyne
1 gram = 980.665 dynes
1 newton = 100,000 dynes
1 kilogram = 1000 grams
1 metric ton = 1000 kilograms

The two systems are related by the following:

1 international pound = 453.59237 grams
1 international foot = 30.48 centimeters

The pound above is the international avoirdupois pound. International units are those adopted by the National Bureau of Standards, effective July 1, 1959.

Units of force are derived from units of mass through the relationship that force is equal to mass times acceleration, which in this case is the acceleration of gravity. The standard acceleration of gravity is 980.665 centimeters per second per second, which is equivalent to 32.1740 international feet per second per second. On this basis, one gram force is equal to one gram mass times 980.665 centimeter per second per second, which equals 980.665 grams mass centimeter per second per second.

The dyne is defined as that force which when applied to one gram mass will produce an acceleration of one centimeter per second per second. The newton is defined as that force which when applied to one kilogram mass will produce an acceleration of one meter per second per second. Therefore, the newton is equivalent to 100,000 dynes. One gram force equals 980.665 dynes.

The poundal when applied to one pound mass will produce an

acceleration of one foot per second per second.   A one pound force when applied to one pound mass will produce an acceleration of 32.1740 feet per second per second.   Therefore, one pound force is equal to 32.1740 poundals.

The units of force may be considered to comprise a basic dimension of measurement and the units of mass derived from force and acceleration.   In such a case, the pound force is considered basic.   By definition, one pound force will produce an acceleration of one foot per second per second when applied to the mass of one slug, which then becomes the basic unit of mass.   The slug equals 32.1740 pounds mass.

The relationship between two units of force, such as the gram and pound, is the same as that between the gram mass and the pound mass.   Any unit of mass may be listed as a unit of force with the same name.   However, only those units of force are listed in the table which are used in the tables on units of energy, power and pressure.

The international and old United States customary systems are related as follows:

1 international unit = 0.9999979 U.S. customary unit
1 U.S. customary unit = 1.0000021 international unit

TABLE 8-1

UNITS OF FORCE	Dyne	Gram	Poundal	Newton	Pound	Kilogram	Short Ton	Metric Ton
Dyne	1	1.019716×10^{-3}	7.233014×10^{-5}	0.00001	2.24809×10^{-6}	1.019716×10^{-6}	1.12404×10^{-9}	1.019716×10^{-9}
Gram	980.665	1	0.0709316	9.80665×10^{-3}	2.2046226×10^{-3}	0.001	1.102311×10^{-6}	0.000001
Poundal	13,825.495	14.0981	1	0.13825495	0.0310810	0.0140981	1.55405×10^{-5}	1.40981×10^{-5}
Newton	100,000	101.9716	7.233014	1	0.224809	0.1019716	1.12404×10^{-4}	1.019716×10^{-4}
Pound	4.44822×10^5	453.59237	32.1740	4.44822	1	0.45359237	0.0005	4.5359237×10^{-4}
Kilogram	980,665	1000	70.9316	9.80665	2.2046226	1	1.102311×10^{-3}	0.001
Short Ton	8.89644×10^8	907,184.74	64,348.1	8896.44	2000	907.18474	1	0.90718474
Metric Ton	9.80665×10^8	1,000,000	70,931.6	9806.65	2204.6226	1000	1.102311	1

9. LENGTH

Units of length listed in the tables are:

INTERNATIONAL

1 inch
1 link = 7.92 inches
1 foot = 12 inches
1 yard = 3 feet
1 fathom = 2 yards
1 rod = 5.5 yards
1 chain = 100 links
1 furlong = 10 chains
1 mile = 1760 yards

METRIC

1 angstrom unit = 0.0001 micron
1 micron = 0.001 millimeter
1 millimeter = 0.1 centimeter
1 centimeter = 0.01 meter
1 meter
1 kilometer = 1000 meters
1 nautical mile = 1852 meters

The two systems of units are related by the following:

1 international inch = 2.54 centimeters
1 international yard = 0.9144 meter

The meter is the fundamental unit of length. The meter is defined as the length equal to 1,650,763.73 wavelengths in a vacuum of the radiation corresponding to the transition between the levels $2p_{10}$ and $5d_5$ of the krypton-86 atom (2).

On July 1, 1959 the international units of length replaced the old United States customary units. The relationship between the metric and United States customary systems was based on the Mendenhall Order of 1893, which stated that one yard was equal to 3600/3937 meter. The meter was then exactly equal to 39.37 inches and the inch was equal to approximately 2.54000508 centimeters, as compared to the international inch being equal to 2.54 centimeters, exactly.

The international and old United States customary systems are related as follows:

1 international unit = 0.999998 U.S. customary unit
(exactly)
1 U.S. customary unit = 1.000002 international unit

On July 1, 1954 the National Bureau of Standards started to use the international nautical mile in lieu of the United States nautical mile. This decision confirmed an official agreement between the Secretary of Commerce and the Secretary of Defense to use the international nautical mile within their respective departments. The international nautical mile, labelled "nautical mile" in these tables, is defined as being equal to 1852 meters. The old United States nautical mile was equal to 1853.248 meters or 6080.20 United States customary feet.

ADDITIONAL UNITS OF LENGTH

INTERNATIONAL

mil = 0.001 inch
printer's point = 0.013837 inch (nearly 1/72 inch)
line (button) = 0.025 inch (1/40 inch)
printer's em (pica) = 0.1666667 inch (1/6 inch)
hand = 4 inches
Gunter's (surveyor's) link = 1 link (in tables) = 7.92 inches
span = 9 inches
Ramden's (engineer's) link = 1 foot
cubit = 18 inches = 1.5 feet
pace = 30 inches = 2.5 feet
ell = 45 inches = 3.75 feet
surveyor's measure rod = 1 rod (in tables) = 16.5 feet
perch = 1 rod (in tables) = 16.5 feet
Gunter's (surveyor's) chain = 1 chain (in tables) = 66 feet
Ramden's (engineer's) chain = 100 feet
bolt (cloth) = 120 feet
skein = 360 feet
cable length = 720 feet
statute mile = 1 mile (in tables) = 5280 feet

U.S. nautical (geographical, sea mile) = 6080.20 feet
statute league = 3 statute miles
nautical league = 3 nautical miles

METRIC

micromicron = 0.000001 micron = 1×10^{-12} meter
millimicron (micromillimeter) = 0.001 micron = 0.000001 millimeter = 1×10^{-9} meter
decimeter = 0.1 meter
dekameter = 10 meters
hectometer = 100 meters
myriameter = 10,000 meters
megameter = 1,000,000 meters

ASTRONOMICAL

The **LIGHT YEAR** is the distance that light will travel in one year. The speed of light is taken as 2.997925×10^{10} centimeters per second (1) and there are 31,536,000 seconds in one calendar year. Therefore, one light year is equal to a distance of 9.454256×10^{17} centimeters, which is equivalent to 9.454256×10^{12} kilometers or 5.874602×10^{12} miles.

The **ASTRONOMICAL UNIT** is the mean distance of the earth from the sun. One astronomical unit equals 92,897,000 miles or 1.4950×10^{8} kilometers.

The **PARSEC** is the distance at which one astronomical unit would subtend an angle of one second. One parsec equals 1.916×10^{13} miles or 3.084×10^{13} kilometers.

TABLE 9-1

UNITS OF LENGTH	Angstrom Unit	Micron	Millimeter	Centimeter	Inch	Link	Foot	Yard
Angstrom Unit	1	0.0001	1×10^{-7}	1×10^{-8}	3.937008×10^{-9}	4.970970×10^{-10}	3.280840×10^{-10}	1.093613×10^{-10}
Micron	10,000	1	0.001	0.0001	3.937008×10^{-5}	4.970970×10^{-6}	3.280840×10^{-6}	1.093613×10^{-6}
Millimeter	1×10^7	1000	1	0.1	0.03937008	4.970970×10^{-3}	3.280840×10^{-3}	1.093613×10^{-3}
Centimeter	1×10^8	10,000	10	1	0.3937008	0.04970970	0.03280840	0.01093613
Inch	2.54×10^8	25,400	25.4	2.54	1	0.1262626	0.08333333	0.02777778
Link	2.01168×10^9	201,168	201.168	20.1168	7.92	1	0.66	0.22
Foot	3.048×10^9	304,800	304.8	30.48	12	1.515152	1	0.3333333
Yard	9.144×10^9	914,400	914.4	91.44	36	4.545455	3	1

(CONTINUED)

TABLE 9-1 (Continued)

UNITS OF LENGTH	Angstrom Unit	Micron	Millimeter	Centimeter	Inch	Link	Foot	Yard
Meter	1×10^{10}	1,000,000	1000	100	39.37008	4.970970	3.280840	1.093613
Fathom	1.8288×10^{10}	1,828,800	1828.8	182.88	72	9.090909	6	2
Rod	5.0292×10^{10}	5,029,200	5029.2	502.92	198	25	16.5	5.5
Chain	2.01168×10^{11}	2.01168×10^{7}	20,116.8	2011.68	792	100	66	22
Furlong	2.01168×10^{12}	2.01168×10^{8}	201,168	20,116.8	7920	1000	660	220
Kilometer	1×10^{13}	1×10^{9}	1,000,000	100,000	39,370.08	4970.970	3280.840	1093.613
Mile	1.609344×10^{13}	1.609344×10^{9}	1,609,344	160,934.4	63,360	8000	5280	1760
Nautical Mile	1.852×10^{13}	1.852×10^{9}	1,852,000	185,200	72,913.39	9206.236	6076.115	2025.372

TABLE 9-2

UNITS OF LENGTH	Meter	Fathom	Rod	Chain	Furlong	Kilometer	Mile	Nautical Mile
Angstrom Unit	1×10^{-10}	5.468066×10^{-11}	1.988388×10^{-11}	4.970970×10^{-12}	4.970970×10^{-13}	1×10^{-13}	6.213712×10^{-14}	5.399568×10^{-14}
Micron	0.000001	5.468066×10^{-7}	1.988388×10^{-7}	4.970970×10^{-8}	4.970970×10^{-9}	1×10^{-9}	6.213712×10^{-10}	5.399568×10^{-10}
Millimeter	0.001	5.468066×10^{-4}	1.988388×10^{-4}	4.970970×10^{-5}	4.970970×10^{-6}	0.000001	6.213712×10^{-7}	5.399568×10^{-7}
Centimeter	0.01	5.468066×10^{-3}	1.988388×10^{-3}	4.970970×10^{-4}	4.970970×10^{-5}	0.00001	6.213712×10^{-6}	5.399568×10^{-6}
Inch	0.0254	0.01388889	5.050505×10^{-3}	1.262626×10^{-3}	1.262626×10^{-4}	2.54×10^{-5}	1.578283×10^{-5}	1.371490×10^{-5}
Link	0.201168	0.11	0.04	0.01	0.001	2.01168×10^{-4}	0.000125	1.086220×10^{-4}
Foot	0.3048	0.1666667	0.06060606	0.01515152	1.515152×10^{-3}	3.048×10^{-4}	1.893939×10^{-4}	1.645788×10^{-4}

(CONTINUED)

TABLE 9-2 (Continued)

UNITS OF LENGTH	Meter	Fathom	Rod	Chain	Furlong	Kilometer	Mile	Nautical Mile
Yard	0.9144	0.5	0.1818182	0.04545455	4.545455×10^{-3}	0.0009144	5.681818×10^{-4}	4.937365×10^{-4}
Meter	1	0.5468066	0.1988388	0.04970970	4.970970×10^{-3}	0.001	6.213712×10^{-4}	5.399568×10^{-4}
Fathom	1.8288	1	0.3636364	0.09090909	9.090909×10^{-3}	1.8288×10^{-3}	1.136364×10^{-3}	9.874730×10^{-4}
Rod	5.0292	2.75	1	0.25	0.025	5.0292×10^{-3}	0.003125	2.715551×10^{-3}
Chain	20.1168	11	4	1	0.1	0.0201168	0.0125	0.01086220
Furlong	201.168	110	40	10	1	0.201168	0.125	0.1086220
Kilometer	1000	546.8066	198.8388	49.70970	4.970970	1	0.6213712	0.5399568
Mile	1609.344	880	320	80	8	1.609344	1	0.8689762
Nautical Mile	1852	1012.686	368.2494	92.06236	9.206236	1.852	1.150779	1

10. MAGNETIC UNITS

There are four tables of magnetic units with each table listing units of both the electromagnetic and electrostatic systems, each of which is based on metric units. Some of the tables also list units in the practical system involving the ampere, the electrical unit of current which is defined in terms of metric units.

Magnetic units listed in the tables are:

MAGNETIC FLUX (FLUX OF MAGNETIC INDUCTION)

$$1 \text{ maxwell} = 1 \text{ line}$$
$$1 \text{ weber} = 1 \times 10^8 \text{ maxwells}$$
$$1 \text{ electrostatic unit} = 2.997925 \times 10^{10} \text{ maxwells}$$

MAGNETIC FLUX DENSITY (MAGNETIC INDUCTION)

$$1 \text{ gauss} = 1 \text{ maxwell per square centimeter}$$
$$1 \text{ tesla} = 10,000 \text{ gausses}$$
$$1 \text{ electrostatic unit} = 2.997925 \times 10^{10} \text{ gausses}$$

MAGNETOMOTIVE FORCE (MAGNETIC POTENTIAL)

$$1 \text{ electrostatic unit} = 3.335640 \times 10^{-11} \text{ gilbert}$$
$$1 \text{ gilbert}$$
$$1 \text{ ampere turn} = 0.4\pi \text{ gilberts}$$

MAGNETIC FIELD INTENSITY

$$1 \text{ electrostatic unit} = 3.335640 \times 10^{-11} \text{ oersted}$$
$$1 \text{ ampere turn per meter} = 0.01 \text{ ampere turn per centimeter}$$
$$1 \text{ oersted}$$
$$1 \text{ ampere turn per centimeter} = 0.4\pi \text{ oersteds}$$

Magnetic flux is the magnetic flow that exists in a magnetic circuit. One line of flux is equal to the maxwell, which is the electromagnetic unit of flux. The weber equals 10^8 maxwells. The basic units of both the electromagnetic and electrostatic systems are related to the metric units through the centimeter and gram mass. The weber relates to the meter and kilogram mass.

Magnetic flux density is the ratio of the flux in any cross-section to the cross-sectional area normal to the direction of the flux. The gauss is equal to a flux density of one maxwell per square centimeter and is the electromagnetic unit of flux density.   The Eleventh General Conference on Weights and Measures adopted on October 14, 1960, the tesla, which is the unit of magnetic flux density in the Système Internationale.   One tesla is equal to one weber per square meter.

Magnetomotive force is that which tends to produce magnetic flux and corresponds to electromotive force, or potential difference, which tends to produce electrical current.   The gilbert is the electromagnetic unit of magnetomotive force.   The ampere turn is a rationalized unit related to the meter and kilogram mass units. One gilbert is equal to $1/0.4\pi$ times the number of ampere turns.

Magnetic field intensity is defined as the vector quantity measured by the force exerted on a unit magnetic pole placed at a point in a vacuum.   The oersted is the electromagnetic unit of field intensity based on the unit magnetic pole which is one that is concentrated at a point and has such strength that, when placed one centimeter away from an exactly similar pole in a medium of unit permeability, the two poles repel each other with the force of one dyne.   The oersted is equal to one gilbert of magnetomotive force per centimeter.

The relationship between the electromagnetic and electrostatic systems is based on the fact that one electrostatic unit of magnetic flux is equal to "c" maxwells, where "c" is the speed of light in a vacuum, 2.997925×10^{10} centimeters per second (1).

ADDITIONAL MAGNETIC UNITS

MAGNETIC FLUX

kiloline = 1000 lines
megaline = 1,000,000 lines
volt second = 1 weber (in tables)
statweber = 1 electrostatic unit (in tables)

MAGNETIC FLUX DENSITY

maxwell per square inch = 0.1550003 maxwell per square
centimeter
line per square inch = 0.1550003 maxwell per square
centimeter
line per square centimeter = 1 maxwell per square centimeter
maxwell per square centimeter = 1 gauss (in tables)

MAGNETOMOTIVE FORCE

abampere turn = 10 ampere turns

MAGNETIC FIELD INTENSITY

ampere turn per inch = 0.3937008 ampere turn per
centimeter
gilbert per centimeter = 1 oers‘ed (in tables)

58

Magnetic Units

TABLE 10-1

UNITS OF MAGNETISM MAGNETIC FLUX	Maxwell	Weber	Electrostatic Unit
Maxwell	1	1×10^{-8}	3.335640×10^{-11}
Weber	1×10^{8}	1	3.335640×10^{-3}
Electrostatic Unit	2.997925×10^{10}	299.7925	1

TABLE 10-2

UNITS OF MAGNETISM MAGNETIC FLUX DENSITY	Gauss	Tesla	Electrostatic Unit
Gauss	1	0.0001	3.335640×10^{-11}
Tesla	10,000	1	3.335640×10^{-7}
Electrostatic Unit	2.997925×10^{10}	2.997925×10^{6}	1

TABLE 10-3

UNITS OF MAGNETISM MAGNETOMOTIVE FORCE	Electrostatic Unit	Gilbert	Ampere Turn
Electrostatic Unit	1	3.335640×10^{-11}	2.654418×10^{-11}
Gilbert	2.997925×10^{10}	1	0.7957747
Ampere Turn	3.767304×10^{10}	1.256637	1

TABLE 10-4

UNITS OF MAGNETISM MAGNETIC FIELD INTENSITY	Electrostatic Unit	Ampere Turn per Meter	Oersted	Ampere Turn per Centimeter
Electrostatic Unit	1	2.654418×10^{-9}	3.335640×10^{-11}	2.654418×10^{-11}
Ampere Turn per Meter	3.767304×10^{8}	1	1.256637×10^{-2}	0.01
Oersted	2.997925×10^{10}	79.57747	1	0.7957747
Ampere Turn per Centimeter	3.767304×10^{10}	100	1.256637	1

11. MASS

Units of mass listed in the tables are:

INTERNATIONAL

1 grain
1 apothecaries scruple = 20 grains
1 troy pennyweight = 24 grains
1 avoirdupois dram = 27.34375 grains
1 apothecaries dram = 3 apothecaries scruples
1 avoirdupois ounce = 16 avoirdupois drams
1 troy ounce = 8 apothecaries drams
1 troy pound = 12 troy ounces
1 avoirdupois pound = 16 avoirdupois ounces
1 short ton = 2000 avoirdupois pounds
1 long ton = 2240 avoirdupois pounds

METRIC

1 milligram = 0.001 gram
1 gram
1 kilogram = 1000 grams
1 metric ton = 1000 kilograms

The two systems are related by the following:

1 international avoirdupois
pound = 453.59237 grams
1 international grain = 0.06479891 gram

The kilogram is the fundamental unit of mass. It is defined as being equal to the mass of the international prototype of the kilogram (2).

On July 1, 1959 the international units of mass replaced the old United States customary units. The relationship between the metric and United States customary systems was based on the Mendenhall Order of 1893, which stated that one avoirdupois pound was equal to 453.5924277 grams. The conversion factor for the international avoirdupois pound was selected so as to be exactly divisible by 7 to yield the exact relationship between the grain and gram.

The international and old United States customary systems are related as follows:

1 international unit = 0.99999987 U.S. customary unit
1 U.S. customary unit = 1.00000013 international unit

ADDITIONAL UNITS OF MASS

INTERNATIONAL

troy dram = 1 apothecaries dram (in tables)
apothecaries ounce = 1 troy ounce (in tables)
apothecaries pound = 1 troy pound (in tables)
slug (geepound) = 32.1740 avoirdupois pounds
short hundredweight (cental,
short quintal) = 100 avoirdupois pounds
long hundredweight (long
quintal) = 112 avoirdupois pounds
short quarter = 500 avoirdupois pounds
long quarter = 560 avoirdupois pounds

METRIC

microgram = 0.000001 gram
centigram = 0.01 gram
decigram = 0.1 gram
metric carat = 0.2 gram
dekagram = 10 grams
hectogram = 100 grams
myriagram = 10,000 grams
metric quintal = 100 kilograms = 100,000 grams
millier (tonne) = 1 metric ton (in tables) = 1,000,000 grams

TABLE 11-1

UNITS OF MASS	Milligram	Grain	Gram	Apothecaries Scruple	Troy Pennyweight	Avoirdupois Dram	Apothecaries Dram	Avoirdupois Ounce
Milligram	1	0.01543236	0.001	7.716179×10^{-4}	6.430149×10^{-4}	5.643834×10^{-4}	2.572060×10^{-4}	3.527396×10^{-5}
Grain	64.79891	1	0.06479891	0.05	0.04166667	0.03657143	0.01666667	2.285714×10^{-3}
Gram	1000	15.43236	1	0.7716179	0.6430149	0.5643834	0.2572060	0.03527396
Apothecaries Scruple	1295.9782	20	1.2959782	1	0.8333333	0.7314286	0.3333333	0.04571429
Troy Pennyweight	1555.17384	24	1.55517384	1.2	1	0.8777143	0.4	0.05485714
Avoirdupois Dram	1771.845	27.34375	1.771845	1.3671875	1.139323	1	0.4557292	0.0625
Apothecaries Dram	3887.9346	60	3.8879346	3	2.5	2.194286	1	0.1371429

(CONTINUED)

TABLE 11-1 (Continued)

UNITS OF MASS	Milligram	Grain	Gram	Apothecaries Scruple	Troy Pennyweight	Avoirdupois Dram	Apothecaries Dram	Avoirdupois Ounce
Avoirdupois Ounce	28,349.523125	437.5	28.349523125	21.875	18.22917	16	7.291667	1
Troy Ounce	31,103.4768	480	31.1034768	24	20	17.55429	8	1.097143
Troy Pound	373,241.7216	5760	373.2417216	288	240	210.6514	96	13.16571
Avoirdupois Pound	453,592.37	7000	453.59237	350	291.6667	256	116.6667	16
Kilogram	1,000,000	15,432.36	1000	771.6179	643.0149	564.3834	257.2060	35.27396
Short Ton	9.0718474×10^8	14,000,000	907,184.74	700,000	583,333.3	512,000	233,333.3	32,000
Metric Ton	1×10^9	1.543236×10^7	1,000,000	771,617.9	643,014.9	564,383.4	257,206.0	35,273.96
Long Ton	1.0160469088×10^9	15,680,000	1,016,046.9088	784,000	653,333.3	573,440	261,333.3	35,840

TABLE 11-2

UNITS OF MASS	Troy Ounce	Troy Pound	Avoirdupois Pound	Kilogram	Short Ton	Metric Ton	Long Ton
Milligram	3.215075×10^{-5}	2.679229×10^{-6}	2.2046226×10^{-6}	0.000001	1.102311×10^{-9}	1×10^{-9}	9.842065×10^{-10}
Grain	2.083333×10^{-3}	1.736111×10^{-4}	1.428571×10^{-4}	6.479891×10^{-5}	7.142857×10^{-8}	6.479891×10^{-8}	6.377551×10^{-8}
Gram	0.03215075	2.679229×10^{-3}	2.2046226×10^{-3}	0.001	1.102311×10^{-6}	0.000001	9.842065×10^{-7}
Apothecaries Scruple	0.04166667	3.472222×10^{-3}	2.857143×10^{-3}	1.2959782×10^{-3}	1.428571×10^{-6}	1.2959782×10^{-6}	1.275510×10^{-6}
Troy Pennyweight	0.05	4.166667×10^{-3}	3.428571×10^{-3}	$1.55517384 \times 10^{-3}$	1.714286×10^{-6}	$1.55517384 \times 10^{-6}$	1.530612×10^{-6}
Avoirdupois Dram	0.05696615	4.747179×10^{-3}	3.90625×10^{-3}	1.771845×10^{-3}	1.953125×10^{-6}	1.771845×10^{-6}	1.743862×10^{-6}
Apothecaries Dram	0.125	0.01041667	8.571429×10^{-3}	3.8879346×10^{-3}	4.285714×10^{-6}	3.8879346×10^{-6}	3.826531×10^{-6}

(CONTINUED)

TABLE 11-2 (Continued)

UNITS OF MASS	Troy Ounce	Troy Pound	Avoirdupois Pound	Kilogram	Short Ton	Metric Ton	Long Ton
Avoirdupois Ounce	0.9114583	0.07595486	0.0625	$2.8349523125 \times 10^{-2}$	3.125×10^{-5}	$2.8349523125 \times 10^{-5}$	2.790179×10^{-5}
Troy Ounce	1	0.08333333	0.06857143	0.0311034768	3.428571×10^{-5}	$3.11034768 \times 10^{-5}$	3.061224×10^{-5}
Troy Pound	12	1	0.8228571	0.3732417216	4.114286×10^{-4}	$3.732417216 \times 10^{-4}$	3.673469×10^{-4}
Avoirdupois Pound	14.58333	1.215278	1	0.45359237	0.0005	4.5359237×10^{-4}	4.464286×10^{-4}
Kilogram	32.15075	2.679229	2.2046226	1	1.102311×10^{-3}	0.001	9.842065×10^{-4}
Short Ton	29,166.67	2430.556	2000	907.18474	1	0.90718474	0.8928571
Metric Ton	32,150.75	2679.229	2204.6226	1000	1.102311	1	0.9842065
Long Ton	32,666.67	2722.222	2240	1016.0469088	1.12	1.0160469088	1

12. POWER

Units of power listed in the tables are:

INTERNATIONAL LENGTH TIMES INTERNATIONAL FORCE PER TIME

1 foot pound per minute
1 foot pound per second = 60 foot pounds per minute
1 horsepower = 550 foot pounds per second

METRIC AND RELATED UNITS (ABSOLUTE)

1 erg per second
1 gram centimeter per second = 980.665 ergs per second
1 kilogram meter per minute = 100,000/60 gram centimeters per second
1 BTU per hour = 1/3600 BTU per second
1 absolute watt = 10,000,000 ergs per second
1 gram calorie per second = 4.1840 absolute watts
1 BTU per minute = 1/60 BTU per second
1 kilogram calorie per minute = 1000/60 gram calories per second
1 metric horsepower = 4500 kilogram meters per minute
1 electrical horsepower = 746 absolute watts
1 absolute kilowatt = 1000 absolute watts
1 BTU per second = 251.9958 gram calories per second

METRIC AND RELATED UNITS (INTERNATIONAL)

1 I.T. BTU per hour = 1/3600 I.T. BTU per second
1 international watt = 1.000165 absolute watts
1 I.T. gram calorie per second = 4.1868 absolute watts
1 I.T. BTU per minute = 1/60 I.T. BTU per second
1 I.T. kilogram calorie per minute = 1000/60 I.T. gram calories per second
1 international kilowatt = 1000 international watts
1 I.T. BTU per second = 251.9958 I.T. gram calories per second

The various systems are related by the following:

1 foot = 30.48 centimeters
1 pound = 453.59237 grams

The foot listed above is the international foot and the pound is the international avoirdupois pound. International units in these cases are those adopted by the National Bureau of Standard, effective July 1, 1959.

Units of force are those derived from units of mass in the cases of the pound, gram and kilogram through the relationship that force is equal to mass times the acceleration of gravity. The standard acceleration of gravity equals 980.665 centimeters per second per second.

The absolute watt is equal to one absolute joule per second by definition. The absolute joule is equivalent to 10,000,000 ergs, and therefore the absolute watt equals 10,000,000 ergs per second.

The gram calorie is defined as being equal to 4.1840 absolute joules. Therefore, one gram calorie per second equals 4.1840 absolute joules per second or 4.1840 absolute watts. The International Steam Table calorie, or I.T. gram calorie, is defined as being equal to 4.1868 absolute joules, and therefore one I.T. gram calorie per second equals 4.1868 absolute watts. The standard gram calorie is also known as the thermochemical calorie, as opposed to the I.T. gram calorie.

Previously, the I.T. gram calorie was taken as equal to 1/860 international watt hour. However, the international watt is equal to 1.000165 absolute watts according to the *National Bureau of Standards Circular C459*, "Announcement of Changes in Electrical and Photometric Units," May 15, 1947. (See ENERGY). On the basis that one I.T. gram calorie equals 1/860 international watt hour, one international watt equals 860/3600 or 0.2388889 I.T. gram calorie per second. On the basis that the international watt equals 1.000165 absolute watts and the I.T. gram calorie per second equals 4.1868 absolute watts, one international watt equals 1.000165/4.1868 or 0.2388853 I.T. gram calorie per second. The latter result is currently correct.

The international watt is related to the metric system of units and is not to be confused with the international system adopted by the National Bureau of Standards, effective July 1, 1959. Furthermore, the international watt no longer has an exact mathematical relationship to the I.T. gram calorie because the latter is currently defined in terms of absolute units of measurement.

The gram calorie was originally defined as the amount of heat required to raise the temperature of one gram of water through one degree Centigrade. The British Thermal Unit, BTU, was originally defined as the amount of heat required to raise the temperature of one pound of water through one degree Fahrenheit.

Although neither of these definitions is exact, the relationship between the British Thermal Unit and gram calorie (and the I.T. BTU and I.T. gram calorie) is still based on comparisons between the pound and gram and the degrees Fahrenheit and Centigrade. It is because the British Thermal Unit is directly related to the gram calorie and through it to the absolute watt, that the British Thermal Unit is grouped in this book with the metric system of units. The I.T. British Thermal Unit is also grouped with the metric units because it is directly related to the I.T. gram calorie and through it to the absolute watt. (See ENERGY).

The horsepower is defined as 550 foot pounds per second. The metric horsepower is defined as the equivalent of 75 kilogram meters per second or 4500 kilogram meters per minute. By definition, one electrical horsepower equals 746 absolute watts.

In cases of the foot pound per minute, foot pound per second, and horsepower, the international and old United States customary units are related as follows:

1 international unit = 0.9999959 U.S. customary unit
1 U.S. customary unit = 1.0000041 international units

The British Thermal Unit based on the international pound is 0.99999987 BTU based on the United States customary pound.

ADDITIONAL UNITS OF POWER

INTERNATIONAL

foot poundal per minute = 0.03108095 foot pound per minute
foot poundal per second = 0.03108095 foot pound per second

METRIC

gram centimeter per minute = 0.00001 kilogram meter per minute
gram calorie per minute = 0.001 kilogram calorie per minute
I.T. gram calorie per minute = 0.001 I.T. kilogram calorie per minute
kilogram meter per second = 100,000 gram centimeters per second

absolute hectowatt = 100 absolute watts
international hectowatt = 100 international watts
kilogram calorie per second = 1000 gram calories per second
I.T. kilogram calorie per second = 1000 I.T. gram calories per second

TABLE 12-1

UNITS OF POWER	Erg per Second	Gram Centimeter per Second	Foot Pound per Minute	Kilogram Meter per Minute	BTU per Hour	I.T. BTU per Hour	Absolute Watt	International Watt
Erg per Second	1	1.019716×10^{-3}	4.425373×10^{-6}	6.11830×10^{-7}	3.414425×10^{-7}	3.412142×10^{-7}	1×10^{-7}	9.99835×10^{-8}
Gram Centimeter per Second	980.665	1	4.339808×10^{-3}	0.0006	3.348407×10^{-4}	3.346168×10^{-4}	9.80665×10^{-5}	9.80503×10^{-5}
Foot Pound per Minute	2.259697×10^{5}	230.4249	1	0.1382550	0.0771556	0.0771040	2.259697×10^{-2}	2.259324×10^{-2}
Kilogram Meter per Minute	1.634442×10^{6}	1666.667	7.23301	1	0.558068	0.557695	0.1634442	0.1634172
BTU per Hour	2.928751×10^{6}	2986.495	12.96081	1.791897	1	0.999331	0.2928751	0.2928268
I.T. BTU per Hour	2.930711×10^{6}	2988.493	12.96949	1.793096	1.000669	1	0.2930711	0.2930227
Absolute Watt	1×10^{7}	10,197.16	44.25373	6.11830	3.414425	3.412142	1	0.999835
International Watt	1.000165×10^{7}	10,198.84	44.26103	6.11931	3.414988	3.412705	1.000165	1
Foot Pound per Second	1.355818×10^{7}	13,825.50	60	8.29530	4.629339	4.626243	1.355818	1.355594
Gram Calorie per Second	4.1840×10^{7}	42,664.93	185.1576	25.59896	14.28595	14.27640	4.1840	4.183310
I.T. Gram Calorie per Second	4.1868×10^{7}	42,693.48	185.2815	25.61609	14.29551	14.28595	4.1868	4.186109

TABLE 12-2

UNITS OF POWER	Erg per Second	Gram Centimeter per Second	Foot Pound per Minute	Kilogram Meter per Minute	BTU per Hour	I.T. BTU per Hour	Absolute Watt	International Watt
BTU per Minute	1.757250×10^8	179,189.7	777.649	107.5138	60	59.9599	17.57250	17.56961
I.T. BTU per Minute	1.758426×10^8	179,309.6	778.169	107.5858	60.0402	60	17.58426	17.58136
Kilogram Calorie per Minute	6.97333×10^8	7.11082×10^5	3085.960	426.6493	238.0992	237.9400	69.7333	69.7218
I.T. Kilogram Calorie per Minute	6.978×10^8	7.11558×10^5	3088.025	426.9348	238.2586	238.0992	69.78	69.7685
Metric Horsepower	7.35499×10^9	7,500,000	32,548.56	4500	2511.305	2509.626	735.499	735.377
Horsepower	7.45700×10^9	7.60402×10^6	33,000	4562.413	2546.136	2544.434	745.700	745.577
Electrical Horsepower	7.46×10^9	7.60708×10^6	33,013.28	4564.250	2547.161	2545.458	746	745.877
Absolute Kilowatt	1×10^{10}	1.019716×10^7	4.425373×10^4	6118.30	3414.425	3412.142	1000	0.999835
International Kilowatt	1.000165×10^{10}	1.019884×10^7	4.426103×10^4	6119.31	3414.988	3412.705	1000.165	1000
BTU per Second	1.054350×10^{10}	1.075138×10^7	4.665893×10^4	6450.83	3600	3597.592	1054.350	1054.176
I.T. BTU per Second	1.055056×10^{10}	1.075858×10^7	4.669016×10^4	6455.15	3602.409	3600	1055.056	1054.882

TABLE 12-3

UNITS OF POWER	Foot Pound per Second	Gram Calorie per Second	I. T. Gram Calorie per Second	BTU per Minute	I.T. BTU per Minute	Kilogram Calorie per Minute	I. T. Kilogram Calorie per Minute
Erg per Second	7.37562×10^{-8}	2.390057×10^{-8}	2.388459×10^{-8}	5.69071×10^{-9}	5.68690×10^{-9}	1.434034×10^{-9}	1.433075×10^{-9}
Gram Centimeter per Second	7.23301×10^{-5}	2.343846×10^{-5}	2.342278×10^{-5}	5.58068×10^{-6}	5.57695×10^{-6}	1.406307×10^{-6}	1.405367×10^{-6}
Foot Pound per Minute	1.666667×10^{-2}	5.40080×10^{-3}	5.39719×10^{-3}	1.285927×10^{-3}	1.285067×10^{-3}	3.240483×10^{-4}	3.238316×10^{-4}
Kilogram Meter per Minute	0.1205502	3.906409×10^{-2}	3.903797×10^{-2}	9.30113×10^{-3}	9.29491×10^{-3}	2.343846×10^{-3}	2.342278×10^{-3}
BTU per Hour	0.2160136	0.0699988	0.0699520	1.666667×10^{-2}	1.665552×10^{-2}	4.199929×10^{-3}	4.197121×10^{-3}
I.T. BTU per Hour	0.2161581	0.0700457	0.0699988	1.667782×10^{-2}	1.666667×10^{-2}	4.202740×10^{-3}	4.199929×10^{-3}
Absolute Watt	0.737562	0.2390057	0.2388459	5.69071×10^{-2}	5.68690×10^{-2}	1.434034×10^{-2}	1.433075×10^{-2}
International Watt	0.737684	0.2390452	0.2388853	5.69165×10^{-2}	5.68784×10^{-2}	1.434271×10^{-2}	1.433312×10^{-2}
Foot Pound per Second	1	0.3240483	0.3238316	7.71556×10^{-2}	7.71040×10^{-2}	1.944290×10^{-2}	1.942989×10^{-2}
Gram Calorie per Second	3.085960	1	0.999331	0.2380992	0.2379400	0.06	0.0599599
I.T. Gram Calorie per Second	3.088025	1.000669	1	0.2382586	0.2380992	0.0600402	0.06

TABLE 12-4

UNITS OF POWER	Foot Pound per Second	Gram Calorie per Second	I.T. Gram Calorie per Second	BTU per Minute	I.T. BTU per Minute	Kilogram Calorie per Minute	I.T. Kilogram Calorie per Minute
BTU per Minute	12.96081	4.199929	4.197121	1	0.999331	0.2519958	0.2518272
I.T. BTU per Minute	12.96949	4.202740	4.199929	1.000669	1	0.2521644	0.2519958
Kilogram Calorie per Minute	51.4327	16.66667	16.65552	3.968321	3.965667	1	0.999331
I.T. Kilogram Calorie per Minute	51.4671	16.67782	16.66667	3.970976	3.968321	1.000669	1
Metric Horsepower	542.476	175.7884	175.6709	41.85509	41.82710	10.54731	10.54025
Horsepower	550	178.2265	178.1074	42.43561	42.40723	10.69359	10.68644
Electrical Horse-power	550.221	178.2983	178.1790	42.45269	42.42429	10.69790	10.69074
Absolute Kilowatt	737.562	239.0057	238.8459	56.9071	56.8690	14.34034	14.33075
International Kilowatt	737.684	239.0452	238.8853	56.9165	56.8784	14.34271	14.33312
BTU per Second	777.649	251.9958	251.8272	60	59.9599	15.11975	15.10963
I.T. BTU per Second	778.169	252.1644	251.9958	60.0402	60	15.12986	15.11975

TABLE 12-5

UNITS OF POWER	Metric Horsepower	Horsepower	Electrical Horsepower	Absolute Kilowatt	International Kilowatt	BTU per Second	I.T. BTU per Second
Erg per Second	1.359622×10^{-10}	1.341022×10^{-10}	1.340483×10^{-10}	1×10^{-10}	9.99835×10^{-11}	9.48451×10^{-11}	9.47817×10^{-11}
Gram Centimeter per Second	1.333333×10^{-7}	1.315093×10^{-7}	1.314564×10^{-7}	9.80665×10^{-8}	9.80503×10^{-8}	9.30113×10^{-8}	9.29491×10^{-8}
Foot Pound per Minute	3.072332×10^{-5}	3.030303×10^{-5}	3.029084×10^{-5}	2.259697×10^{-5}	2.259324×10^{-5}	2.143212×10^{-5}	2.141779×10^{-5}
Kilogram Meter per Minute	2.222222×10^{-4}	2.191822×10^{-4}	2.190941×10^{-4}	1.634442×10^{-4}	1.634172×10^{-4}	1.550189×10^{-4}	1.549152×10^{-4}
BTU per Hour	3.981993×10^{-4}	3.927519×10^{-4}	3.925939×10^{-4}	2.928751×10^{-4}	2.928268×10^{-4}	2.777778×10^{-4}	2.775920×10^{-4}
I.T BTU per Hour	3.984658×10^{-4}	3.930148×10^{-4}	3.928567×10^{-4}	2.930711×10^{-4}	2.930227×10^{-4}	2.779637×10^{-4}	2.777778×10^{-4}
Absolute Watt	1.359622×10^{-3}	1.341022×10^{-3}	1.340483×10^{-3}	0.001	9.99835×10^{-4}	9.48451×10^{-4}	9.47817×10^{-4}
International Watt	1.359846×10^{-3}	1.341243×10^{-3}	1.340704×10^{-3}	1.000165×10^{-3}	0.001	9.48608×10^{-4}	9.47974×10^{-4}
Foot Pound per Second	1.843399×10^{-3}	1.818182×10^{-3}	1.817450×10^{-3}	1.355815×10^{-3}	1.355594×10^{-3}	1.285927×10^{-3}	1.285067×10^{-3}
Gram Calorie per Second	5.68866×10^{-3}	5.61084×10^{-3}	5.60858×10^{-3}	4.1840×10^{-3}	4.18310×10^{-3}	3.968321×10^{-3}	3.965667×10^{-3}
I.T. Gram Calorie per Second	5.69246×10^{-3}	5.61459×10^{-3}	5.61233×10^{-3}	4.1868×10^{-3}	4.186109×10^{-3}	3.970976×10^{-3}	3.968321×10^{-3}

TABLE 12-6

UNITS OF POWER	Metric Horsepower	Horsepower	Electrical Horsepower	Absolute Kilowatt	International Kilowatt	BTU per Second	I.T. BTU per Second
BTU per Minute	2.389196×10^{-2}	2.356512×10^{-2}	2.355564×10^{-2}	1.757250×10^{-2}	1.756961×10^{-2}	1.666667×10^{-2}	1.665552×10^{-2}
I.T. BTU per Minute	2.390795×10^{-2}	2.358089×10^{-2}	2.357140×10^{-2}	1.758426×10^{-2}	1.758136×10^{-2}	1.667782×10^{-2}	1.666667×10^{-2}
Kilogram Calorie per Minute	9.48109×10^{-2}	9.35139×10^{-2}	9.34763×10^{-2}	6.97333×10^{-2}	6.97218×10^{-2}	6.61387×10^{-2}	6.60944×10^{-2}
I.T. Kilogram Calorie per Minute	9.48744×10^{-2}	9.35765×10^{-2}	9.35389×10^{-2}	0.06978	6.97685×10^{-2}	6.61829×10^{-2}	6.61387×10^{-2}
Metric Horsepower	1	0.986320	0.985923	0.735499	0.735377	0.697585	0.697118
Horsepower	1.013870	1	0.999598	0.745700	0.745577	0.707260	0.706787
Electrical Horsepower	1.014278	1.000402	1	0.746	0.745877	0.707545	0.707072
Absolute Kilowatt	1.359622	1.341022	1.340483	1	0.999835	0.948451	0.947817
International Kilowatt	1.359846	1.341243	1.340704	1.000165	1	0.948608	0.947974
BTU per Second	1.433517	1.413907	1.413338	1.054350	1.054176	1	0.999331
I.T. BTU per Second	1.434477	1.414853	1.414284	1.055056	1.054882	1.000669	1

13. PRESSURE

Units of pressure listed in the tables are:

INTERNATIONAL FORCE PER INTERNATIONAL AREA

1 pound per square foot
1 pound per square inch = 144 pounds per square foot
1 short ton per square foot = 2000 pounds per square foot

METRIC FORCE PER METRIC AREA

1 dyne per square centimeter
1 kilogram per square meter = 0.1 gram per square centimeter
1 gram per square centimeter = 980.665 dynes per square
centimeter
1 millibar = 1000 dynes per square centimeter
1 metric ton per square meter = 100 grams per square centimeter
1 newton per square centimeter = 100,000 dynes per square
centimeter
1 kilogram per square centimeter = 1000 grams per square centimeter
1 bar = 1000 millibars
1 atmosphere = 1,013,250 dynes per square
centimeter

BASED ON COLUMN OF MERCURY AT 0°C

1 millimeter of mercury at 0°C
1 inch of mercury at 0°C = 25.4 millimeters of mercury
at 0°C

BASED ON COLUMN OF WATER AT 4°C

1 inch of water at 4°C
1 foot of water at 4°C = 12 inches of water at 4°C

The various systems are related by the following:

1 inch = 25.4 millimeters
1 square inch = 6.4516 square centimeters
1 pound = 453.59237 grams

density of mercury at $0°C$ = 13.5951 grams per cubic
centimeter
density of water at $4°C$ = 0.999972 gram per cubic
centimeter

The inch and foot listed above are the international inch and
foot, the pound is the international avoirdupois pound, and the
short ton is the international short ton.   International units in these
cases are those adopted by the National Bureau of Standards, effective
July 1, 1959.

Units of force are those derived from units of mass in the cases
of gravity.   The standard acceleration of gravity equals 980.665
centimeters per second per second.   The dyne is equal to one gram
of gravity.   The standard acceleration of gravity equals 980.665 centimeters per second per second.   The dyne is equal to one gram
mass centimeter per second per second.

The liter is now equivalent to 1000 cubic centimeters; previously
the liter was defined as a unit of capacity equal to the volume occupied
by one kilogram of pure water at its maximum density (at a temperature of $4°$ C, practically) and under the standard atmospheric pressure
of 760 millimeters of mercury.   Therefore, the density of water at
$4°$ C, practically, was previously equal to one kilogram per liter,
and also one gram per milliliter.   The milliliter was previously
considered equal to 1.000028 cubic centimeters, and the density of
water at $4°$ C, practically, was and is equal to 0.999972 gram per
cubic centimeter, not exactly.   It is no longer correct to consider
the density of water at $4°$ C equal to one gram per milliliter on the
basis that the milliliter is now equal to one cubic centimeter, and the
density of water may now be considered equal to 0.999972 gram
per cubic centimeter or milliliter.

With the density of water considered equal to 0.999972 gram
mass per cubic centimeter, a one centimeter cube of water will exert
a pressure of 0.999972 gram force on its base of one square centimeter.
A column of water one inch in height will exert a pressure of 2.54
times 0.999972 or 2.539929 grams force on a base of one square
centimeter.   Conversion to international units may be made using
the relationships that one pound equals 453.59237 grams and one
square inch equals 6.4516 square centimeters.

The atmosphere is defined, according to Reference 1, as equal
to 1,013,250 dynes per square centimeter.   Historically, the atmosphere was defined as the equivalent of 760 millimeters of mercury
at $0°$ C.   The two definitions are in very good agreement.   The
density of mercury at $0°$ C is 13.5951 grams mass per cubic centimeter.

A column of mercury 760 millimeters in height will exert a pressure 76 times 13.5951 or 1033.23 grams force on a base of one square centimeter. One gram force is equal to 980.665 dynes. Therefore, one atmosphere is equivalent to 76 times 13.5951 times 980.665, which equals a fraction more than 1,013,250 dynes per square centimeter. If the value 1,013,250 dynes per square centimeter equals one atmosphere is taken as exact, the height of the column of mercury is calculated to equal 759.99989 millimeters, which may be rounded off to six significant figures as 760.000 millimeters. In the tables, both values are presented as being exact equivalents of the atmosphere: 1,013,250 dynes per square centimeter and 760 millimeters of mercury at 0° C.

The international and old United States customary units of pressure are related as follows:

1 international unit = 1.0000019 U.S. customary units
1 U.S. customary unit = 0.9999981 international unit

The above relationships refer to the pound per square foot, pound per square inch, and the short ton per square foot. The inch of mercury, inch of water, and foot of water have the following relationships:

1 international unit = 0.999998 U.S. customary unit
(exactly)
1 U.S. customary unit = 1.000002 international units

ADDITIONAL UNITS OF PRESSURE

INTERNATIONAL

poundal per square foot = 0.03108095 pound per square foot
ounce per square inch = 0.0625 pound per square inch
= 9 pounds per square foot
long ton per square foot = 2240 pounds per square foot
short ton per square inch = 2000 pounds per square inch
= 288,000 pounds per square foot
long ton per square inch = 2240 pounds per square inch
= 322,560 pounds per square foot

METRIC

barye = 1 dyne per square centimeter (in tables)

newton per square meter = 10 dynes per square centimeter

kilogram per square millimeter = 100 kilograms per square centimeter

COLUMN OF MERCURY AT 0°C

centimeter of mercury = 10 millimeters of mercury

COLUMN OF WATER AT 4°C

centimeter of water = 0.3937008 inch of water

TABLE 13-1

UNITS OF PRESSURE	Dyne per Square Centimeter	Kilogram per Square Meter	Pound per Square Foot	Gram per Square Centimeter	Millibar	Millimeter of Mercury at 0 °C	Inch of Water at 4 °C	Foot of Water at 4 °C
Dyne per Square Centimeter	1	1.019716×10^{-2}	2.088543×10^{-3}	1.019716×10^{-3}	0.001	7.50062×10^{-4}	4.014743×10^{-4}	3.345619×10^{-5}
Kilogram per Square Meter	98.0665	1	0.2048161	0.1	0.0980665	7.35559×10^{-2}	3.937118×10^{-2}	3.280932×10^{-3}
Pound per Square Foot	478.8026	4.882428	1	0.4882428	0.4788026	0.359131	0.1922269	1.601891×10^{-2}
Gram per Square Centimeter	980.665	10	2.048161	1	0.980665	0.735559	0.3937118	3.280932×10^{-2}
Millibar	1000	10.19716	2.088543	1.019716	1	0.750062	0.4014743	3.345619×10^{-2}
Millimeter of Mercury at 0 °C	1333.22	13.5951	2.78450	1.35951	1.33322	1	0.535255	4.46046×10^{-2}
Inch of Water at 4 °C	2490.819	25.39929	5.20218	2.539929	2.490819	1.86827	1	8.333333×10^{-2}
Foot of Water at 4 °C	29,889.83	304.7915	62.4262	30.47915	29.88983	22.4192	12	1

TABLE 13-2

UNITS OF PRESSURE	Dyne per Square Centimeter	Kilogram per Square Meter	Pound per Square Foot	Gram per Square Centimeter	Millibar	Millimeter of Mercury at 0 °C	Inch of Water at 4 °C	Foot of Water at 4 °C
Inch of Mercury at 0 °C	33,863.9	345.316	70.7262	34.5316	33.8639	25.4	13.5955	1.13296
Pound per Square Inch	68,947.6	703.070	144	70.3070	68.9476	51.7149	27.68068	2.306723
Metric Ton per Square Meter	98,066.5	1000	204.8161	100	98.0665	73.5559	39.37118	3.280932
Newton per Square Centimeter	100,000	1019.716	208.8543	101.9716	100	75.0062	40.14743	3.345619
Short Ton per Square Foot	9.57605×10^{5}	9764.86	2000	976.486	957.605	718.263	384.4539	32.03782
Kilogram per Square Centimeter	980,665	10,000	2048.161	1000	980.665	735.559	393.7118	32.80932
Bar	1,000,000	10,197.16	2088.543	1019.716	1000	750.062	401.4743	33.45619
Atmosphere	1,013,250	10,332.27	2116.217	1033.227	1013.250	760	406.7939	33.89949

TABLE 13-3

UNITS OF PRESSURE	Inch of Mercury at 0°C	Pound per Square Inch	Metric Ton per Square Meter	Newton per Square Centimeter	Short Ton per Square Foot	Kilogram per Square Centimeter	Bar	Atmosphere
Dyne per Square Centimeter	2.95300×10^{-5}	1.450377×10^{-5}	1.019716×10^{-5}	0.00001	1.044272×10^{-6}	1.019716×10^{-6}	0.000001	9.86923×10^{-7}
Kilogram per Square Meter	2.89590×10^{-3}	1.422334×10^{-3}	0.001	9.80665×10^{-4}	1.024081×10^{-4}	0.0001	9.80665×10^{-5}	9.67841×10^{-5}
Pound per Square Foot	1.41390×10^{-2}	6.944444×10^{-3}	4.882428×10^{-3}	4.788026×10^{-3}	0.0005	4.882428×10^{-4}	4.788026×10^{-4}	4.725414×10^{-4}
Gram per Square Centimeter	2.89590×10^{-2}	1.422334×10^{-2}	0.01	9.80665×10^{-3}	1.024081×10^{-3}	0.001	9.80665×10^{-4}	9.67841×10^{-4}
Millibar	2.95300×10^{-2}	1.450377×10^{-2}	1.019716×10^{-2}	0.01	1.044272×10^{-3}	1.019716×10^{-3}	0.001	9.86923×10^{-4}
Millimeter of Mercury at 0°C	3.937008×10^{-2}	1.93368×10^{-2}	1.35951×10^{-2}	1.33322×10^{-2}	1.39225×10^{-3}	1.35951×10^{-3}	1.33322×10^{-3}	1.31579×10^{-3}
Inch of Water at 4°C	7.35539×10^{-2}	3.612628×10^{-2}	2.539929×10^{-2}	2.490819×10^{-2}	2.601092×10^{-3}	2.539929×10^{-3}	2.490819×10^{-3}	2.458248×10^{-3}
Foot of Water at 4°C	0.882646	0.4335154	0.3047915	0.2988983	3.121311×10^{-2}	3.047915×10^{-2}	2.988983×10^{-2}	2.949897×10^{-2}

TABLE 13-4

UNITS OF PRESSURE	Inch of Mercury at 0 °C	Pound per Square Inch	Metric Ton per Square Meter	Newton per Square Centimeter	Short Ton per Square Foot	Kilogram per Square Centimeter	Bar	Atmosphere
Inch of Mercury at 0 °C	1	0.491154	0.345316	0.338639	3.53631×10^{-2}	3.45316×10^{-2}	3.38639×10^{-2}	3.34211×10^{-2}
Pound per Square Inch	2.03602	1	0.703070	0.689476	0.072	7.03070×10^{-2}	6.89476×10^{-2}	6.80460×10^{-2}
Metric Ton per Square Meter	2.89590	1.422334	1	0.980665	0.1024081	0.1	9.80665×10^{-2}	9.67841×10^{-2}
Newton per Square Centimeter	2.95300	1.450377	1.019716	1	0.1044272	0.1019716	0.1	9.86923×10^{-2}
Short Ton per Square Foot	28.2781	13.88889	9.76486	9.57605	1	0.976486	0.957605	0.945083
Kilogram per Square Centimeter	28.9590	14.22334	10	9.80665	1.024081	1	0.980665	0.967841
Bar	29.5300	14.50377	10.19716	10	1.044272	1.019716	1	0.986923
Atmosphere	29.9213	14.69595	10.33227	10.13250	1.058108	1.033227	1.013250	1

14. TIME

Units of time listed in the table are:

$$
\begin{aligned}
1 \text{ second} \\
1 \text{ minute} &= 60 \text{ seconds} \\
1 \text{ hour} &= 60 \text{ minutes} \\
1 \text{ day} &= 24 \text{ hours} \\
1 \text{ week} &= 7 \text{ days} \\
1 \text{ 30-day month} &= 720 \text{ hours} \\
1 \text{ 31-day month} &= 744 \text{ hours} \\
1 \text{ year} &= 365 \text{ days}
\end{aligned}
$$

These units are those used in common practice rather than units of time based on astronomical exactness. For example, the year in the table is defined as exactly 365 days, which is equivalent to 31,536,000 seconds.

The second is the fundamental unit of time. The second is defined as the duration of 9,192,631,770 periods of the radiation corresponding to the transition between the two hyperfine levels of the ground state of the cesium-133 atom (2).

Until recently, the second was defined as 1/86,400 of a mean solar day. However, the second of mean solar time did not represent an invariable standard because the average rotational speed of the earth may vary by almost one part in ten million. In addition, seasonal changes in the sidereal rotation of the earth amounts to about 1 part in 100 million within the year. The orbital motion of the earth about the sun and that of the moon about the earth behave more regularly and furnish the basis for uniform time, called Ephemeris Time. The Eleventh General Conference on Weights and Measures on October 14, 1960, confirmed the action of the International Committee on Weights and Measures in defining the second as equal to 1/31,556,925.9747 of the tropical year 1900. This number precisely defines the second which is identical with the second of Ephemeris Time.

The definition of the second based on the cesium-133 atom was authorized at 1725 Paris time, October 8, 1964, by the Twelfth General Conference of Weights and Measures. The new definition facilitates the expression of the results of high precision time and frequency measurements.

ADDITIONAL UNITS OF TIME

28 day month = 2,419,200 seconds = 40,320 minutes = 672 hours
= 4 weeks
29 day month = 2,505,600 second = 41,760 minutes = 696 hours
= 4.142857 weeks
366 day year = 31,622,400 seconds = 527,040 minutes = 8784 hours
= 52.28571 weeks

The solar day is the time of one revolution of the earth on its axis as measured by the interval between two successive transits of the sun over the same meridian. Because solar days are of unequal duration, the mean solar day represents the average.

The sidereal day is the time of one revolution of the earth on its axis relative to the position of the earth with respect to the stars.

The solar or tropical year is the time for the earth to complete one revolution around the sun based on two consecutive returns of the sun to the vernal equinox.

The sidereal year is the time for the earth to complete one revolution around the sun based on the position of the earth with respect to a given star—the time for the earth to return to the same position with respect to the same star.

tropical year = 365.24219879 mean solar days−(6.14 ×
10^{-8} days) (T−1900)
sidereal year = 365.25636042 mean solar days + (1.1 ×
10^{-9} days) (T−1900)
(In both cases, T is the number of the year, as 1960)

The uncorrected values, where T is 1900 and the correction factors equal zero, are:

tropical year = 365.24219879 mean solar days
= 31,556,925.975 mean solar seconds
= 365 days, 5 hours, 48 minutes, 45.975 seconds (mean solar time)
sideral year = 365.25636042 mean solar days
= 31,558,149.540 mean solar seconds
= 365 days, 6 hours, 9 minutes, 9.540 seconds (mean solar time)

Abbreviated conversion factors where correction factors have no effect are:

mean solar (tropical) year = 365.2422 days (mean solar time)
= 8765.8128 hours (mean solar time)
= 365 days, 5 hours, 48 minutes, 46 seconds (mean solar time)

sidereal year = 365.2564 days (mean solar time)
= 8766.144 hours (mean solar time)
= 365 days, 6 hours, 9 minutes, 13 seconds (based on 365.2564 mean solar days)
= 365 days, 6 hours, 8 minutes, 38 seconds (based on 8766.144 mean solar hours)

TABLE 14-1

UNITS OF TIME	Second	Minute	Hour	Day	Week	30 Day Month	31 Day Month	Year
Second	1	0.01666667	2.777778×10^{-4}	1.157407×10^{-5}	1.653439×10^{-6}	3.858025×10^{-7}	3.733572×10^{-7}	3.170979×10^{-8}
Minute	60	1	0.01666667	6.944444×10^{-4}	9.920635×10^{-5}	2.314815×10^{-5}	2.240143×10^{-5}	1.902588×10^{-6}
Hour	3600	60	1	0.04166667	5.952381×10^{-3}	1.388889×10^{-3}	1.344086×10^{-3}	1.141553×10^{-4}
Day	86,400	1440	24	1	0.1428571	0.03333333	0.03225806	2.739726×10^{-3}
Week	604,800	10,080	168	7	1	0.2333333	0.2258065	0.01917808
30 Day Month	2,592,000	43,200	720	30	4.285714	1	0.9677419	0.08219178
31 Day Month	2,678,400	44,640	744	31	4.428571	1.033333	1	0.08493151
Year	31,536,000	525,600	8760	365	52.14286	...	...	1

15. VELOCITY

Units of velocity listed in the tables are:

INTERNATIONAL LENGTH PER TIME

1 foot per minute
1 foot per second = 60 feet per minute
1 mile per hour = 88 feet per minute
1 mile per minute = 5280 feet per minute
1 mile per second = 5280 feet per second

METRIC LENGTH PER TIME

1 centimeter per second = 0.6 meter per minute
1 meter per minute = 0.06 kilometer per hour
1 kilometer per hour
1 knot = 1.852 kilometers per hour
1 meter per second = 60 meters per minute
1 kilometer per minute = 60 kilometers per hour

The two systems of units are related by the following:

1 international foot = 30.48 centimeters

On July 1, 1959, the international units of length replaced the old United States customary units, which were based on the Mendenhall Order of 1893. The knot listed above is defined as one international nautical mile per hour. The international nautical mile, actually a metric system unit rather than an international system unit, was adopted by the National Bureau of Standards on July 1, 1954, as the replacement of the United States nautical mile of 1853.248 meters. The international nautical mile is defined as being equal to 1852 meters. Therefore, one knot is equal to 1852 meters per hour or 1.852 kilometers per hour.

The international and old United States customary. systems are related as follows:

1 international unit = 0.999998 U.S. customary unit
(exactly)
1 U.S. customary unit = 1.000002 international unit

ADDITIONAL UNITS OF VELOCITY

INTERNATIONAL

inch per hour = 1.388889×10^{-3} foot per minute

foot per hour = 0.01666667 foot per minute

yard per hour = 0.05 foot per minute

inch per minute = 0.08333333 foot per minute

yard per minute = 3 feet per minute

inch per second = 0.08333333 foot per second

yard per second = 3 feet per second

METRIC

centimeter per minute = 0.01 meter per minute

meter per hour = 0.001 kilometer per hour

kilometer per second = 1000 meters per second

TABLE 15-1

UNITS OF VELOCITY	Foot per Minute	Centimeter per Second	Meter per Minute	Kilometer per Hour	Foot per Second	Mile per Hour
Foot per Minute	1	0.508	0.3048	0.018288	0.01666667	0.01136364
Centimeter per Second	1.968504	1	0.6	0.036	0.03280840	0.02236936
Meter per Minute	3.280840	1.666667	1	0.06	0.05468066	0.03728227
Kilometer per Hour	54.68066	27.77778	16.66667	1	0.9113444	0.6213712
Foot per Second	60	30.48	18.288	1.09728	1	0.6818182
Mile per Hour	88	44.704	26.8224	1.609344	1.466667	1
Knot	101.2686	51.44444	30.86667	1.852	1.687810	1.150779
Meter per second	196.8504	100	60	3.6	3.280840	2.236936
Kilometer per Minute	3280.840	1666.667	1000	60	54.68066	37.28227
Mile per Minute	5280	2682.24	1609.344	96.56064	88	60
Mile per Second	316.800	160,934.4	96,560.64	5793.6384	5280	3600

TABLE 15-2

UNITS OF VELOCITY	Knot	Meter per second	Kilometer per Minute	Mile per Minute	Mile per Second
Foot per Minute	9.874730×10^{-3}	0.00508	0.0003048	1.893939×10^{-4}	3.156566×10^{-6}
Centimeter per Second	0.01943844	0.01	0.0006	3.728227×10^{-4}	6.213712×10^{-6}
Meter per Minute	0.03239741	0.01666667	0.001	6.213712×10^{-4}	1.035619×10^{-5}
Kilometer per Hour	0.5399568	0.2777778	0.01666667	0.01035619	1.726031×10^{-4}
Foot per Second	0.5924838	0.3048	0.018288	0.01136364	1.893939×10^{-4}
Mile per Hour	0.8689762	0.44704	0.0268224	0.01666667	2.777778×10^{-4}
Knot	1	0.5144444	0.03086667	0.01917966	3.196610×10^{-4}
Meter per Second	1.943844	1	0.06	0.03728227	6.213712×10^{-4}
Kilometer per Minute	32.39741	16.66667	1	0.6213712	0.01035619
Mile per Minute	52.13857	26.8224	1.609344	1	0.01666667
Mile per Second	3128.314	1609.344	96.56064	60	1

16. VOLUME

Units of volume, including liquid and dry measures, listed in the tables are:

INTERNATIONAL VOLUME

1 cubic inch
1 cubic foot = 1728 cubic inches
1 cubic yard = 27 cubic feet

METRIC VOLUME

1 cubic centimeter = 0.000001 cubic meter
1 cubic meter

U.S. LIQUID MEASURE

1 minim
1 fluid dram = 60 minims
1 fluid ounce = 8 fluid drams
1 gill = 4 fluid ounces
1 liquid pint = 4 gills
1 liquid quart = 2 liquid pints
1 gallon = 4 liquid quarts
1 petroleum barrel = 42 gallons

U.S. DRY MEASURE

1 dry pint
1 dry quart = 2 dry pints
1 peck = 8 dry quarts
1 bushel = 4 pecks

METRIC CAPACITY

1 milliliter = 0.001 liter
1 liter
1 kiloliter = 1000 liters

The various systems are related by the following:

1 international cubic inch = 16.387064 cubic centimeters
1 U.S. gallon = 231 international cubic inches
1 U.S. bushel = 2150.42 international cubic
inches
1 liter = 1000 cubic centimeters

The international units of volume are based on the international units of length adopted by the National Bureau of Standards, effective July 1, 1959.   Units of United States liquid and dry measures were originally defined in terms of the old United States customary units but are now defined in international units.

The liter is defined as equal to one cubic decimeter. Thus the liter is equal to 1000 cubic centimeters and the milliliter equals one cubic centimeter. This is a change from the previous definition of the liter as a unit of capacity equal to the volume occupied by one kilogram of pure water at its maximum density (at a temperature of 4°C, practically) and under the standard atmospheric pressure of 760 millimeters of mercury.

The following paragraph is taken from an article titled, "Actions Taken by the Twelfth General Conference on Weights and Measures", *National Bureau of Standards Technical News Bulletin*, page 207, December, 1964:

> "The liter, defined up to now as the volume occupied by one kilogram of water, differs from a cubic decimeter by about 28 millionths, and this discrepancy —slightly out of line with other international measurements—has frequently caused difficulty in precision work. The Conference therefore abrogated the old definition, and made the liter merely a special name for the cubic decimeter. The resolution in which this action was taken, however, pointed out that the word 'liter' should not be used to express the results of volume measurements of high precision."

According to the old definition, one liter equalled 1000.028 cubic centimeters. Thus the liter as previously defined was 28 parts per million larger than the liter as presently defined.

The international and old United States customary units of volume are related as follows:

1 international unit = 0.999994 U.S. customary unit
1 U.S. customary unit = 1.000006 international unit

ADDITIONAL UNITS OF VOLUME

INTERNATIONAL VOLUME

board foot × 1 foot × 1 inch = 144 cubic inches

cord foot = 4 feet × 4 feet × 1 foot = 16 cubic feet

cord = 4 feet × 4 feet × 8 feet = 128 cubic feet

acre inch = 6,272,640 cubic inches = 3630 cubic feet

acre foot = 43,560 cubic feet

METRIC VOLUME

cubic milliliter = 0.001 cubic centimeter

cubic decimeter = 0.001 cubic meter

decistere = 0.1 cubic meter

stere = 1 cubic meter (in tables)

dekastere = 10 cubic meters

cubic dekameter = 1000 cubic meters

cubic hectometer = 1,000,000 cubic meters

cubic kilometer = 1×10^9 cubic meters

U.S. LIQUID MEASURE

firkin = 9 gallons

liquid barrel = 31.5 gallons

hogshead = 63 gallons

tun = 252 gallons

One gallon is equal to a cube 6.135792 inches on a side.

U.S. DRY MEASURE

barrel for cranberries = 5826 cubic inches

barrel for fruits, vegetables, and other dry commodities, other than cranberries = 7056 cubic inches = 105 dry quarts

chaldron = 36 bushels

One bushel is equal to a cube 12.90747 inches on a side.

METRIC CAPACITY

microliter = 0.000001 liter
centiliter = 0.01 liter
deciliter = 0.1 liter
dekaliter = 10 liters
hectoliter = 100 liters

TABLE 16-1

UNITS OF VOLUME	Cubic Centimeter	Milliliter	Cubic Inch	Liter	Cubic Foot	Cubic Yard	Cubic Meter	Kiloliter
Cubic Centimeter	1	1	6.102374×10^{-2}	0.001	3.531467×10^{-5}	1.307951×10^{-6}	0.000001	0.000001
Milliliter	1	1	6.102374×10^{-2}	0.001	3.531467×10^{-5}	1.307951×10^{-6}	0.000001	0.000001
Cubic Inch	16.387064	16.387064	1	1.6387064×10^{-2}	5.787037×10^{-4}	2.143347×10^{-5}	1.6387064×10^{-5}	1.6387064×10^{-5}
Liter	1000	1000	61.02374	1	3.531467×10^{-2}	1.307951×10^{-3}	0.001	0.001
Cubic Foot	28,316.85	28,316.85	1728	28.31685	1	3.703704×10^{-2}	2.831685×10^{-2}	2.831685×10^{-2}
Cubic Yard	764,554.9	764,554.9	46,656	764.5549	27	1	0.7645549	0.7645549
Cubic Meter	1,000,000	1,000,000	61,023.74	1000	35.31467	1.307951	1	1
Kiloliter	1,000,000	1,000,000	61,023.74	1000	35.31467	1.307951	1	1

TABLE 16-2

UNITS OF VOLUME LIQUID MEASURE	Minim	Fluid Dram	Fluid Ounce	Gill	Liquid Pint	Liquid Quart	Gallon	Petroleum Barrel
Minim	1	1.666667×10^{-2}	2.083333×10^{-3}	5.208333×10^{-4}	1.302083×10^{-4}	6.510417×10^{-5}	1.627604×10^{-5}	3.875248×10^{-7}
Fluid Dram	60	1	0.125	0.03125	7.8125×10^{-3}	3.90625×10^{-3}	9.765625×10^{-4}	2.325149×10^{-5}
Fluid Ounce	480	8	1	0.25	0.0625	0.03125	7.8125×10^{-3}	1.860119×10^{-4}
Gill	1920	32	4	1	0.25	0.125	0.03125	7.440476×10^{-4}
Liquid Pint	7680	128	16	4	1	0.5	0.125	2.976190×10^{-3}
Liquid Quart	15,360	256	32	8	2	1	0.25	5.952381×10^{-3}
Gallon	61,440	1024	128	32	8	4	1	2.380952×10^{-2}
Petroleum Barrel	2,580,480	43,008	5376	1344	336	168	42	1

TABLE 16-3

UNITS OF VOLUME DRY MEASURE	Dry Pint	Dry Quart	Peck	Bushel
Dry Pint	1	0.5	0.0625	0.015625
Dry Quart	2	1	0.125	0.03125
Peck	16	8	1	0.25
Bushel	64	32	4	1

TABLE 16-4

UNITS OF VOLUME VOLUME VS. LIQUID MEASURE	Minim	Fluid Dram	Fluid Ounce	Gill	Liquid Pint	Liquid Quart	Gallon	Petroleum Barrel
Cubic Centimeter	16.23073	0.2705122	3.381402×10^{-2}	8.453506×10^{-3}	2.113376×10^{-3}	1.056688×10^{-3}	2.641721×10^{-4}	6.289811×10^{-6}
Milliliter	16.23073	0.2705122	3.381402×10^{-2}	8.453506×10^{-3}	2.113376×10^{-3}	1.056688×10^{-3}	2.641721×10^{-4}	6.289811×10^{-6}
Cubic Inch	265.9740	4.432900	0.5541126	0.1385281	3.463203×10^{-2}	1.731602×10^{-2}	4.329004×10^{-3}	1.030715×10^{-4}
Liter	16,230.73	270.5122	33.81402	8.453506	2.113376	1.056688	0.2641721	6.289811×10^{-3}
Cubic Foot	459,603.1	7660.052	957.5065	239.3766	59.84416	29.92208	7.480519	0.1781076
Cubic Yard	1.240928×10^{7}	206,821.4	25,852.68	6463.169	1615.792	807.8961	201.9740	4.808905
Cubic Meter	1.623073×10^{7}	270,512.2	33,814.02	8453.506	2113.376	1056.688	264.1721	6.289811
Kiloliter	1.623073×10^{7}	270,512.2	33,814.02	8453.506	2113.376	1056.688	264.1721	6.289811

TABLE 16-5

UNITS OF VOLUME LIQUID MEASURE VS. VOLUME	Cubic Centimeter	Milliliter	Cubic Inch	Liter	Cubic Foot	Cubic Yard	Cubic Meter	Kiloliter
Minim	6.161152×10^{-2}	6.161152×10^{-2}	3.759766×10^{-3}	6.161152×10^{-5}	2.175790×10^{-6}	8.058483×10^{-8}	6.161152×10^{-8}	6.161152×10^{-8}
Fluid Dram	3.696691	3.696691	0.2255859	3.696691×10^{-3}	1.305474×10^{-4}	4.835090×10^{-6}	3.696691×10^{-6}	3.696691×10^{-6}
Fluid Ounce	29.57353	29.57353	1.8046875	2.957353×10^{-2}	1.044379×10^{-3}	3.868072×10^{-5}	2.957353×10^{-5}	2.957353×10^{-5}
Gill	118.2941	118.2941	7.21875	0.1182941	4.177517×10^{-3}	1.547229×10^{-4}	1.182941×10^{-4}	1.182941×10^{-4}
Liquid Pint	473.1765	473.1765	28.875	0.4731765	1.671007×10^{-2}	6.188915×10^{-4}	4.731765×10^{-4}	4.731765×10^{-4}
Liquid Quart	946.3529	946.3529	57.75	0.9463529	3.342014×10^{-2}	1.237783×10^{-3}	9.463529×10^{-4}	9.463529×10^{-4}
Gallon	3785.41178	3785.41178	231	3.78541178	0.1336806	4.951132×10^{-3}	$3.78541178 \times 10^{-3}$	$3.78541178 \times 10^{-3}$
Petroleum Barrel	158,987.3	158,987.3	9702	158.9873	5.614583	0.2079475	0.1589873	0.1589873

TABLE 16-6

UNITS OF VOLUME VOLUME VS. DRY MEASURE	Dry Pint	Dry Quart	Peck	Bushel
Cubic Centimeter	1.816166×10^{-3}	9.080830×10^{-4}	1.135104×10^{-4}	2.837759×10^{-5}
Milliliter	1.816166×10^{-3}	9.080830×10^{-4}	1.135104×10^{-4}	2.837759×10^{-5}
Cubic Inch	2.976163×10^{-2}	1.488081×10^{-2}	1.860102×10^{-3}	4.650254×10^{-4}
Liter	1.816166	0.9080830	0.1135104	2.837759×10^{-2}
Cubic Foot	51.42809	25.71405	3.214256	0.8035640
Cubic Yard	1388.559	694.2793	86.78491	21.69623
Cubic Meter	1816.166	908.0830	113.5104	28.37759
Kiloliter	1816.166	908.0830	113.5104	28.37759

TABLE 16-7

UNITS OF VOLUME DRY MEASURE VS. VOLUME	Cubic Centimeter	Milliliter	Cubic Inch	Liter	Cubic Foot	Cubic Yard	Cubic Meter	Kiloliter
Dry Pint	550.6105	550.6105	33.6003125	0.5506105	1.944463×10^{-2}	7.201713×10^{-4}	5.506105×10^{-4}	5.506105×10^{-4}
Dry Quart	1101.221	1101.221	67.200625	1.101221	3.888925×10^{-2}	1.440343×10^{-3}	1.101221×10^{-3}	1.101221×10^{-3}
Peck	8809.768	8809.768	537.605	8.809768	0.3111140	1.152274×10^{-2}	8.809768×10^{-3}	8.809768×10^{-3}
Bushel	35,239.07	35,239.07	2150.42	35.23907	1.244456	4.609096×10^{-2}	3.523907×10^{-2}	3.523907×10^{-2}

17. REFERENCES

(1) NEW VALUES FOR THE PHYSICAL CONSTANTS.

Recommended by the National Academy of Sciences - National Research Council. National Bureau of Standards Technical News Bulletin, 175–177 (October, 1963).

(2) NEW DEFINITIONS AUTHORIZED FOR SI BASE UNITS.

(SI is the abbreviation for Système International or the International System of Units.)
National Bureau of Standards Technical News Bulletin, 34 (February, 1969).

The following were used as general references:

1. National Bureau of Standards Technical News Bulletin, 43, No. 1 (January, 1959).
2. National Bureau of Standards Technical News Bulletin, 44, No. 12 (December, 1960).
3. L.V. Judson, "Units of Weight and Measure, Definitions and Tables of Equivalents," National Bureau of Standards Miscellaneous Publication 214 (July 1, 1955).
4. L.V. Judson, National Bureau of Standards Miscellaneous Publication 233 (December 20, 1960).

18. APPENDIX A
SYSTEMS OF UNITS

TABLE 18-1

	CENTIMETER GRAM MASS SECOND	METER KILOGRAM MASS SECOND (Système Internationale)	FOOT POUND MASS SECOND	FOOT POUND FORCE SECOND
LENGTH	centimeter	meter	foot	foot
MASS	gram	kilogram	pound	$\left(\dfrac{\text{slug}}{\dfrac{\text{pound force}}{\text{foot per sec}^2}}\right)$
TIME	second	second	second	second
FORCE	$\left(\dfrac{\text{dyne}}{\dfrac{\text{gram centimeter}}{\text{second}^2}}\right)$	$\left(\dfrac{\text{newton}}{\dfrac{\text{kilogram meter}}{\text{second}^2}}\right)$	$\left(\dfrac{\text{poundal}}{\dfrac{\text{pound mass foot}}{\text{second}^2}}\right)$	pound
AREA	square centimeter	square meter	square foot	square foot
VOLUME	cubic centimeter	cubic meter	cubic foot	cubic foot
DENSITY AND CONCEN-TRATION	$\dfrac{\text{gram}}{\text{cubic centimeter}}$	$\dfrac{\text{kilogram}}{\text{cubic meter}}$	$\dfrac{\text{pound}}{\text{cubic foot}}$	$\dfrac{\text{slug}}{\text{cubic foot}}$
VELOCITY	$\dfrac{\text{centimeter}}{\text{second}}$	$\dfrac{\text{meter}}{\text{second}}$	$\dfrac{\text{foot}}{\text{second}}$	$\dfrac{\text{foot}}{\text{second}}$
FLOW	$\dfrac{\text{cubic centimeter}}{\text{second}}$	$\dfrac{\text{cubic meter}}{\text{second}}$	$\dfrac{\text{cubic foot}}{\text{second}}$	$\dfrac{\text{cubic foot}}{\text{second}}$
PRESSURE	$\dfrac{\text{dyne}}{\text{square centimeter}}$	$\dfrac{\text{newton}}{\text{square meter}}$	$\dfrac{\text{poundal}}{\text{square foot}}$	$\dfrac{\text{pound}}{\text{square foot}}$
ENERGY	erg (dyne centimeter)	joule (newton meter)	foot poundal	foot pound
POWER	$\dfrac{\text{erg}}{\text{second}}$	watt (joule/second)	$\dfrac{\text{foot poundal}}{\text{second}}$	$\dfrac{\text{foot pound}}{\text{second}}$

TABLE 18-2

| | CENTIMETER GRAM MASS SECOND | | METER KILOGRAM MASS SECOND |
	Electrostatic Units	*Electromagnetic Units*	*(Système Internationale)*
ELECTRICAL CHARGE	statcoulomb	abcoulomb	coulomb
ELECTRICAL CURRENT	statampere	abampere	ampere
ELECTRICAL RESISTANCE	statohm	abohm	ohm
ELECTRICAL POTENTIAL	statvolt	abvolt	volt
ELECTRICAL CAPACITANCE	statfarad	abfarad	farad
ELECTRICAL INDUCTANCE	stathenry	abhenry	henry
ELECTRICAL CONDUCTANCE	statmho	abmho	mho
MAGNETIC FLUX	electrostatic unit	maxwell	weber
MAGNETIC FLUX DENSITY	electrostatic unit	gauss	tesla
MAGNETIC POTENCIAL	electrostatic unit	gilbert	ampere turn
MAGNETIC FIELD INTENSITY	electrostatic unit	oersted	ampere turn per meter

19. APPENDIX B

COMPARISON OF INTERNATIONAL AND U.S. CUSTOMARY UNITS

In this appendix typical examples are given to demonstrate the comparisons between the international system of units adopted by the National Bureau of Standards, effective July 1, 1959 and the old United States customary units. For this purpose, the following basic conversion factors and abbreviations are used:

1 international pound (int. lb) = 453.59237 grams (gm)
1 international inch (int. in.) = 2.54 centimeters (cm)
1 international foot (int. ft) = 12 international inches
1 U.S. customary pound (U.S. lb) = 453.5924277 grams
1 centimeter = 0.3937 U.S. customary inch (U.S. in)
1 U.S. customary foot (U.S. ft) = 12 U.S. customary inches

1. Length

$$\frac{2.54 \text{ cm}}{\text{int. in.}} \times \frac{0.3937 \text{ U.S. in.}}{\text{cm}} = 0.999998 \text{ (exactly)}$$

1 int. in. = 0.999998 U.S. in.
1 U.S. in. = 1.000002 int. in.

2. Mass

$$\frac{453.59237 \text{ gm}}{\text{int. lb}} \times \frac{\text{U.S. lb}}{453.5924277 \text{ gm}} = 0.99999987$$

1 int. lb = 0.99999987 U.S. lb
1 U.S. lb = 1.00000013 int. lb.

3. Area

$$\frac{(2.54 \text{ cm})^2}{(\text{int. in.})^2} \times \frac{(0.3937 \text{ U.S. in.})^2}{\text{cm}^2} = 0.999996$$

1 int. square in. = 0.999996 U.S. square in.
1 U.S. square in. = 1.000004 int. square in.

4. Volume

$$\frac{(2.54 \text{ cm})^3}{(\text{int. in.})^3} \times \frac{(0.3937 \text{ U.S. in.})^3}{\text{cm}^3} = 0.999994$$

1 int. cubic in. = 0.999994 U.S. cubic in.
1 U.S. cubic in. = 1.000006 int. cubic in.

5. Velocity

Velocity has the dimensions of length per unit time. Therefore, the relationships are the same as those for units of length.

1 int. ft per sec = 0.999998 U.S. ft per sec
1 U.S. ft per sec = 1.000002 int. ft per sec

6. Flow

Flow has the dimensions of volume per unit time. The relationships are the same as those for units of volume.

1 int. cubic ft per sec = 0.999994 U.S. cubic ft per sec
1 U.S. cubic ft per sec = 1.000006 int. cubic ft per sec

7. Density and Concentration

Density and concentration have the dimensions of mass per unit volume.

$$\frac{453.59237 \text{ gm}}{\text{int. lb}} \times \frac{(\text{int. in.})^3}{(2.54 \text{ cm})^3} \times \frac{\text{U.S. lb}}{453.5924277 \text{ gm}} \times \frac{\text{cm}^3}{(0.3937 \text{ U.S. in.})^3}$$

$$= 1.00000587$$

1 int. lb per cubic in. = 1.00000587 U.S. lb per cubic in.
1 U.S. lb per cubic in. = 0.99999413 int. lb per cubic in.

8. Force

Force has the dimensions of mass times length per time squared.

$$\frac{453.59237 \text{ gm}}{\text{int. lb}} \times \frac{2.54 \text{ cm}}{\text{int. in.}} \times \frac{\text{U.S. lb}}{453.5924277 \text{ gm}} \times \frac{0.3937 \text{ U.S. in.}}{\text{cm}}$$

$$= 0.9999979$$

1 int. lb mass ft per sec^2 = 0.9999979 U.S. lb mass ft per sec^2
1 U.S. lb mass ft per sec^2 = 1.0000021 int. lb mass ft per sec^2

9. Pressure

Pressure has the dimensions of force per unit area, which reduces to mass times length per length squared times time squared, ML/L^2T^2, which further reduces to M/LT^2, mass per length time squared.

$$\frac{453.59237\ gm}{int.\ lb} \times \frac{int.\ in.}{2.54\ cm} \times \frac{U.S.\ lb}{453.5924277\ gm} \times \frac{cm}{0.3937\ U.S.\ in.}$$

$$= 1.0000019$$

1 int. lb force per square in. = 1.0000019 U.S.lb force per square in.

1 U.S. lb force per square in. = 0.9999981 int. lb force per square in.

10. Energy

Energy has the dimensions of force times length, which reduces to mass times length squared per time squared, ML^2/T^2.

$$\frac{453.59237\ gm}{int.\ lb} \times \frac{(2.54\ cm)^2}{(int.\ in.)^2} \times \frac{U.S.\ lb}{453.5924277\ gm} \times \frac{(0.3937\ U.S.\ in.)^2}{cm^2}$$

$$= 0.9999959$$

1 int. lb force ft = 0.9999959 U.S. lb force ft
1 U.S. lb force ft = 1.0000041 int. lb force ft

11. Power

Power has the dimensions of energy per unit time and therefore the relationships are the same as those for energy.

1 int. lb force ft per sec = 0.9999959 U.S. lb force ft per sec
1 U.S. lb force ft per sec = 1.0000041 int. lb force ft per sec

INDEX